Lecture Notes in Computer Science 11525

Commenced Publication in 1973
Founding and Former Series Editors:
Gerhard Goos, Juris Hartmanis, and Jan van Leeuwen

More information about this series at http://www.springer.com/series/7407

Alonso Castillo-Ramirez ·
Pedro P. B. de Oliveira (Eds.)

Cellular Automata and Discrete Complex Systems

25th IFIP WG 1.5 International Workshop, AUTOMATA 2019
Guadalajara, Mexico, June 26–28, 2019
Proceedings

 Springer

Editors
Alonso Castillo-Ramirez (iD)
University of Guadalajara
Guadalajara, Mexico

Pedro P. B. de Oliveira (iD)
Mackenzie Presbyterian University
Sao Paulo, Brazil

ISSN 0302-9743   ISSN 1611-3349 (electronic)
Lecture Notes in Computer Science
ISBN 978-3-030-20980-3   ISBN 978-3-030-20981-0 (eBook)
https://doi.org/10.1007/978-3-030-20981-0

LNCS Sublibrary: SL1 – Theoretical Computer Science and General Issues

This Springer imprint is published by the registered company Springer Nature Switzerland AG
The registered company address is: Gewerbestrasse 11, 6330 Cham, Switzerland

Preface

This volume contains the full papers presented at the 25th International Workshop on Cellular Automata (CA) and Discrete Complex Systems (DCS), AUTOMATA 2019, held during June 26–28, 2019, at the University Center of Exact Sciences and Engineering, CUCEI, University of Guadalajara, México. The conference was the official annual event of the International Federation for Information Processing (IFIP), Working Group 5, on CA and DCS, of the Technical Committee 1, on Foundations of Computer Science.

AUTOMATA 2019 is part of an annual series of conferences established in 1995 as a collaboration forum between researchers in CA and DCS. Current topics of the conference include, but are not limited to, dynamical, topological, ergodic, and algebraic aspects of CA and DCS, algorithmic and complexity issues, emergent properties, formal languages, symbolic dynamics, tilings, models of parallelism and distributed systems, timing schemes, synchronous versus asynchronous models, phenomenological descriptions, scientific modeling, and practical applications.

AUTOMATA 2019 was the fourth time that the conference took place in a Latin American country, the last time having been in Santiago, Chile, in 2011. As a celebration of its Silver Jubilee, the event had a special session commemorating the date. We had three invited talks given by Pablo Arrighi, Hector Zenil, and Tullio Ceccherini-Silberstein; we sincerely thank them for accepting the invitation and their very interesting presentations. The abstracts of the invited talks are included here.

We received ten submissions as full papers for the conference. Each submission was reviewed by three members of the Program Committee. Based on these reviews and an open discussion, seven papers were accepted to be presented at the conference and to be included in the proceedings. We thank all authors for their contributions and hard work that made this event possible.

The conference program also involved short presentations of exploratory papers that are not included in these proceedings, and we wish to extend our thanks to the authors of the exploratory submissions.

We are indebted to the Steering Committee, Program Committee, and additional reviewers for their valuable help during the last months. We acknowledge the generous funding provided by the IFIP and the University of Guadalajara toward the organization of this event.

We are very grateful for the support of the local Organizing Committee and the authorities of CUCEI; in particular, our sincere thanks to Dr. Humberto Gutiérrez Pulido, Head of the Department of Mathematics, and to Dr. Ruth Padilla Muñoz, Rector of CUCEI. Finally, we acknowledge the excellent cooperation from the *Lecture Notes in Computer Science* team of Springer for their help in producing this volume in time for the conference.

April 2019

Alonso Castillo-Ramirez
Pedro P. B. de Oliveira

Organization

Steering Committee

Enrico Formenti   Université Côte d'Azur, France
Jan Baetens   Ghent University, Belgium
Jarkko Kari   University of Turku, Finland
Pedro de Oliveira   Universidade Presbiteriana Mackenzie, Brazil
Turlough Neary   University of Zurich and ETH Zurich, Switzerland

Program Committee

Alberto Dennunzio   Università degli Studi di Milano-Bicocca, Italy
Alonso Castillo-Ramirez   Universidad de Guadalajara, México
  (Co-chair)
Andrew Adamatzky   University of the West of England, UK
Edgardo Ugalde   Universidad Autónoma de San Luis Potosí, México
Enrico Formenti   Université Côte d'Azur, France
Guillaume Theyssier   CNRS and Aix Marseille Université, France
Hector Zenil   SciLifeLab and Karolinska Institute, Sweden
Henryk Fukś   Brock University, Canada
Ilkka Törmä   University of Turku, Finland
Jan Baetens   Ghent University, Belgium
Janko Gravner   University of California Davis, USA
Jarkko Kari   University of Turku, Finland
Katsunobu Imai   Hiroshima University, Japan
Luciano Margara   University of Bologna, Italy
Martin Kutrib   Justus-Liebig-Universität Gießen, Germany
Maximilien Gadouleau   Durham University, UK
Nazim Fatès   Inria Nancy, France
Paola Flocchini   University of Ottawa, Canada
Pedro de Oliveira (Co-chair)   Universidade Presbiteriana Mackenzie, Brazil
Pietro Di Lena   University of Bologna, Italy
Reem Yassawi   Université Claude Bernard Lyon 1, France
Siamak Taati   University of Groningen, The Netherlands
Thomas Worsch   Karlsruhe Institute of Technology, Germany
Turlough Neary   University of Zurich and ETH Zurich, Switzerland

Additional Reviewers

Kévin Perrot
Luca Manzoni
Antonio E. Porreca
Ville Salo
Simon Beier

Local Organizing Committee

Alonso Castillo-Ramirez
Humberto Gutiérrez Pulido
Osbaldo Mata Gutiérrez
María de la Paz Suárez Fernández
Juan Manuel Márquez Bobadilla

Abstracts of Invited Talks

Quantum Cellular Automata: Computability and Universality

P. Arrighi

Aix-Marseille University, Université de Toulon, CNRS, LIS, Marseille,
and IXXI, Lyon, France
pablo.arrighi@univ-amu.fr

This talk will provide an overview of the field of Quantum Cellular Automata (QCA). QCA consist in an array of identical finite-dimensional quantum systems. The whole array evolves in discrete time steps by iterating a linear operator G. Moreover this global evolution G is shift-invariant (it acts everywhere in the same way), causal (information cannot be transmitted faster than some fixed number of cells per time step), and unitary (the condition required by the postulate of evolutions in quantum theory, akin to reversibility). I will motivate their studies through the distinction between finite-dimensional quantum evolutions (unitary matrices, a.k.a quantum automata) and infinite-dimensional quantum evolutions (unitary operators, a.k.a operators). It turns out that basic questions surrounding computability and universality had only been solved for quantum automata, and not for quantum operators. We addressed the question of their computability in [1], building upon a structural theorem that actually decomposes the axiomatic version of QCA [9, 17] into an infinitely repeating quantum circuit of local gates [7, 8], which we will recall. We also addressed the question of the universality of quantum operators in a series of works [2–4] which we will summarize, and relate to the fascinating question of how much particle physics can be recast in terms of QCA [6, 10, 13, 18]. In all of these results, quantum information flows on a fixed, classical background: a rigid grid. Could the background itself be quantum? We will touch on this rather topical question [5, 11, 16].

References

1. Arrighi, P., Dowek, G.: The physical church-turing thesis and the principles of quantum theory. Int. J. Found. Comput. Sci. **23** (2012)
2. Arrighi, P., Grattage, J.: A quantum game of life. In: Second Symposium on Cellular Automata "Journées Automates Cellulaires" (JAC 2010), Turku, December 2010, TUCS Lecture Notes, vol. 13, pp. 31–42 (2010)
3. Arrighi, P., Grattage, J.: Intrinsically universal n-dimensional quantum cellular automata. J. Comput. Syst. Sci. **78**, 1883–1898 (2012)
4. Arrighi, P., Grattage, J.: Partitioned quantum cellular automata are intrinsically universal. Nat. Comput. **11**, 13–22 (2012)
5. Arrighi, P., Martiel, S.: Quantum causal graph dynamics. Phys. Rev. D. **96**(2), 024026, (2017). Pre-print, arXiv:1607.06700

6. Arrighi, P., Nesme, V., Forets, M.: The dirac equation as a quantum walk: higher dimensions, observational convergence. J. Phys. A: Math. Theor. **47**(46), 465302 (2014)
7. Arrighi, P., Nesme, V., Werner, R.: Unitarity plus causality implies localizability. J. Comput. Syst. Sci. **77**, 372–378 (2010). QIP 2010 (long talk)
8. Arrighi, P., Nesme, V., Werner, R.: Unitarity plus causality implies localizability (full version). J. Comput. Syst. Sci. **77**(2), 372–378 (2011)
9. Arrighi, P., Nesme, V.; Werner, R.F.: One-dimensional quantum cellular automata over finite, unbounded configurations. In: Martín-Vide, C., Otto, F., Fernau, H. (eds.) LATA 2008. LNCS, vol. 5196, pp. 64–75. Springer, Heidelberg (2008)
10. Bialynicki-Birula, I.: Weyl, Dirac, and Maxwell equations on a lattice as unitary cellular automata. Phys. Rev. D. **49**, 6920–6927 (1994)
11. Chiribella, G., D'Ariano, G.M., Perinotti, P., Valiron, B.: Quantum computations without definite causal structure. Phys. Rev. A. **88**(2), 022318 (2013)
12. Gromov, M.: Endomorphisms of symbolic algebraic varieties. J. Eur. Math. Soc. **1**(2), 109–197 (1999)
13. Meyer, D.A.: From quantum cellular automata to quantum lattice gases. J. Stat. Phys. **85**(5–6), 551–574 (1996)
14. Moore, E.F.: Machine models of self-reproduction. In: In Mathematical Problems in Biological Sciences, Proceedings of Symposia in Applied Mathematics. American Mathematical Society (1962)
15. Myhill, J.: The converse of Moore's Garden-of-Eden theorem. Proc. Am. Math. Soc. 685–686 (1963, in press)
16. Oreshkov, O., Costa, F., Brukner, Č.: Quantum correlations with no causal order. Nat. Commun. **3**, 1092 (2012)
17. Schumacher, B., Werner, R.: Reversible quantum cellular automata (2004). arXiv pre-print, quant-ph/0405174
18. Succi, S., Benzi, R.: Lattice boltzmann equation for quantum mechanics. Physica D: Nonlinear Phenomena **69**(3), 327–332 (1993)

Algorithmic Information Dynamics Reconstructs the Space-Time and Phase-Space Dynamics of Cellular Automata

Hector Zenil[1,2,3], Narsis A. Kiani[2,3], Jesper Tegnér[3,4,5]

[1] Oxford Immune Algorithmics, Oxford, UK
[2] Algorithmic Dynamics Lab, Karolinska Institute, Karolinska University
Hospital, Stockholm, Sweden
hector.zenil@algorithmicnaturelab.org
https://www.hectorzenil.net/
[3] Unit of Computational Medicine, Center for Molecular Medicine,
Karolinska Institute, Stockholm, Sweden
{hector.zenil,narsis.kiani,jesper.tegner}@ki.se
http://www.compmed.se/
[4] BESE Division, King Abdullah University of Science
and Technology (KAUST), Kingdom of Saudi Arabia
[5] Algorithmic Nature Group, LABORES for the Natural
and Digital Sciences, Paris, France
https://algorithmicnature.org/

Current widely used tools for causal inference are mostly based on classical statistics, including machine learning, and draw heavily upon statistical patterns in data, leading to inverse problems that involve estimating parameters such as initial conditions and boundary conditions on the basis of observed data. These inverse problems are often ill posed, not only because correlation does not imply causation, but also more importantly, because causation does not imply correlation, on which most of our current data-driven inference techniques are based. This means that in most, if not all physical and biological instances, it is not only difficult to infer stable states from noisy and chaotic data, but also that such inferences are impossible to make even in principle, when using traditional statistically-based tools to find true generating mechanisms. Here we show how the new theory of Algorithmic Information Dynamics [1, 2] can help reconstruct dynamical systems in application to one of the most studied discrete computational model, cellular automata. We will show how the theory of algorithmic probability can find reconstructions of space-time evolutions and phase-space landscapes of these representative discrete dynamical systems.

At the core of Algorithmic Information Dynamics [1, 2], the algorithmic causal calculus that we introduced, is the quantification of the change of complexity of a system under natural or induced perturbations, particularly the direction (sign) and magnitude of the difference of algorithmic information approximations denoted by C between an object G, such a cellular automaton or a graph and its mutated version G', e.g. the flip of a cell bit (or a set of bits) or the removal of an edge e from G (denoted by $G \backslash e = G'$). The difference $|C(G) - C(G \backslash e)|$ is an estimation of the shared algorithmic mutual information of G and $G\backslash e$. If e does not contribute to the description of G, then $|C(G) - C(G\backslash e)| \leq log_2 |G|$, where $|G|$ is the uncompressed size of G, i.e. the difference will be very small and at most a function of the graph size, and thus $C(G)$ and $C(G\backslash e)$ have almost the same complexity. If, however, $|C(G) - C(G\backslash e)| \leq log_2 |G|$ bits, then G and $G\backslash e$ share at least n bits of algorithmic information in element e, and the removal of e results in a loss of information. In contrast, if $C(G) - C(G\backslash e) > n$, then e cannot be explained by G alone, nor is it algorithmically not contained/derived from G, and it is therefore a fundamental part of the description of G, with e as a generative causal mechanism in G, or else it is not part of G but has to be explained independently, e.g. as noise. Whether it is noise or part of the generating mechanism of G depends on the relative magnitude of n with respect to $C(G)$ and to the original causal content of G itself. If G is random, then the effect of e will be small in either case, but if G is richly causal and has a very small generating program, then e as noise will have a greater impact on G than would removing e from an already short description of G. However, if $|C(G) - C(G\backslash e)| \leq log_2 |G|$, where $|G|$ is e.g. the vertex count of a graph, or the runtime of a cellular automaton, G, then e is contained in the algorithmic description of G and can be recovered from G itself (e.g. by running the program from a previous step until it produces G with e from $G\backslash e$).

We show how we can infer and reconstruct space-time evolutions by quantification of the disruptiveness of a perturbation. We can then extract the generating mechanism from the ordered time indices, from least to most disruptive and produce candidate generating models. Simpler rules have simpler hypotheses, with an almost perfect correspondence in row order. Some systems may look more disordered than others, but locally the relationship between single rows is mostly preserved (indicating local reversibility).

We show that the later in time a perturbation is injected into a dynamical system the less it contributes to the algorithmic information content of the overall space-time evolution. We then move from discrete 2D systems to the reconstruction of phase spaces and space-time evolutions of N-dimensional, continuous, chaotic, incomplete and even noisy dynamical systems.

References

1. Zenil, H., et al.: An Algorithmic Information Calculus for Causal Discovery and Reprogramming Systems, bioaRXiv (2018). https://doi.org/10.1101/185637
2. Zenil, H., Kiani, N.A., Tegnér, J.: Algorithmic information dynamics of emergent, persistent, and colliding particles in the game of life. In: Adamatzky, A. (ed.) From Parallel to Emergent Computing, pp. 367–383. Taylor & Francis/CRC Press (2019)

The Garden of Eden Theorem: From Cellular Automata to Algebraic Dynamical Systems

Tullio Ceccherini-Silberstein

Dipartimento di Ingegneria, Università del Sannio, 82100 Benevento, Italy
tullio.cs@sbai.uniroma1.it

The *Garden of Eden theorem* proved by Moore [5] and Myhill [6] in 1963 is a central result in the Theory of Cellular Automata and Symbolic Dynamics. Let A be a finite set, called the *alphabet*, and let G be a group, called the *universe*. The *configuration space* is $A^G := \{x : G \to A\} \equiv \prod_{g \in G} A$ equipped with the *prodiscrete topology* (the product topology, where each factor A is discrete) and the G-action defined by $[gx](h) = x(g^{-1}h)$ for all $g, h \in G$ and $x \in A^G$, called the *G-shift*. A *cellular automaton* is a map $\tau : A^G \to A^G$ which is *continuous* and *G-equivariant* (that is, $\tau(gx) = g\tau(x)$ for all $g \in G$ and $x \in A^G$). Two configurations $x, y \in A^G$ are said to be *almost equal* if there exists a finite subset $F \subset G$ such that $x(g) = y(g)$ for all $g \in G \backslash F$. Clearly, almost equality is an equivalence relation. One then says that a map $\tau : A^G \to A^G$ is *pre-injective* if its restriction to each almost equality class is injective. We are now in position to state the Garden of Eden theorem of Moore and Myhill.

Theorem 1 (Garden of Eden theorem of Moore and Myhill (1963)).
Let A be a finite set, let $G = \mathbb{Z}^d$ (the free Abelian group of rank $d \geq 1$), and let $\tau : A^G \to A^G$ be a cellular automaton. Then τ is surjective if and only if it is pre-injective.

A group G is *amenable* if it admits a finitely additive left-invariant probability measure on the set $\mathcal{P}(G)$ of all subsets of G, that is, a map $\mu : \mathcal{P}(G) \to [0, 1]$ such that $\mu(A \cup B) = \mu(A) + \mu(B) - \mu(A \cap B)$, $\mu(gA) = \mu(A)$ (here $gA = \{ga : a \in A\} \subset G$) for all $A, B \subset G$ and $g \in G$, and $\mu(G) = 1$. The class of amenable group contains all finite groups, all Abelian groups, and more generally all solvable groups, and all finitely generated groups of sub-exponential growth, and it is closed under the operations of taking: subgroups, quotients, extensions, and direct limits. On the other hand, the free group $\mathbb{F}_2$ of rank 2 (and therefore every group containing a non-Abelian free subgroup) is not amenable. In 1999 we proved the following:

Theorem 2 (Garden of Eden theorem for amenable groups [3], see also [1]).
Let A be a finite set, let G be an amenable group, and let $\tau : A^G \to A^G$ be a cellular automaton. Then τ is surjective if and only if it is pre-injective.

Gromov [4] suggested that the Garden of Eden theorem could be extended to dynamical systems with a suitable hyperbolic flavor.

A *dynamical system* is a pair (X, G) where X is a compact metrizable space, called the *phase space* and G is a *countable* group acting on X by homeomorphisms. A continuous map $\tau : X \to X$ which is G-equivariant (that is, commuting with the action of G) is called an *endomorphism* of the dynamical system (X, G). One says that two points $x, y \in X$ are *homoclinic* if for every $\varepsilon > 0$ there exists a finite subset $F \subset G$ such that $d(gx, gy) < \varepsilon$ for all $g \in G \backslash F$. Homoclinicity is an equivalence relation: a map $\tau : X \to X$ which is injective on each homoclinicity class is called *pre-injectice*. One says that (X, G) satisfies the Garden of Eden theorem if every endomorphism is surjective if and only if it is pre-injective.

Example 1. Let A be a finite set and let G be a countable group. Then (A^G, G), where G acts on the configuration space A^G by the G-shift. Moreover $x, y \in A^G$ are homoclinic if an only if they are almost equal. As a consequence, the two notions of pre-injectivity coincide. Finally, the endomorphisms of (A^G, G) are precisely the cellular automata.

Let now G be a countable group and denote by $\mathbb{Z}[G]$ the *integral group ring* of G, that is, the ring consisting of all *finite* sums $\sum_{g \in G} a_g \mathbf{g}$ with $a_g \in \mathbb{Z}$. Let $f \in \mathbb{Z}[G]$ and denote by $M_f := \mathbb{Z}[G]/f\mathbb{Z}[G]$ the quotient of $\mathbb{Z}[G]$ by the principal ideal generated by f. Note that M_f is a $\mathbb{Z}[G]$-module. Its *Pontryagin dual* $X_f := \widehat{M_f}$ is a compact metrizable Abelian group which is also a $\mathbb{Z}[G]$-module, and therefore, in particular, bears an action of G. The dynamical system (X_f, G) is called the *principal algebraic dynamical system* associated with the *polynomial* $f \in \mathbb{Z}[G]$ (see [7]). In 2018 we proved the following:

Theorem 3 (Garden of Eden theorem for Harmonic Models [2]).
Let G be a countable Abelian group not virtually $\mathbb{Z}^d$ for $d = 1, 2$. Suppose that $f = \sum_g f_g \mathbf{g} \in \mathbb{Z}[G]$ satisfies the following conditions: (i) $\sum_{g \in G} f_g = 0$, (ii) $f_g \leq 0$ for all $g \in G \backslash \{1_G\}$, (iii) $f_g = f_{g^{-1}}$ for all $g \in G$ (i.e., f is self-adjoint), (iv) supp $(f) := \{g \in G : f_g \neq 0\}$, the support of f, generates the group G, and (v) f is primitive, that is, there is no integer $m \geq 2$ dividing all coefficients of f. Then (X_f, G) satisfies the Garden of Eden Theorem.

References

1. Ceccherini-Silberstein, T., Coornaert, M.: Cellular Automata and Groups, Springer Monographs in Mathematics. Springer-Verlag, Berlin (2010)
2. Ceccherini-Silberstein, T., Coornaert, M., Li, H.: Homoclinically expansive actions and a Garden of Eden theorem for harmonic models. Comm. Math. Phys. (to appear)
3. Ceccherini-Silberstein, T., Mach, A., Scarabotti, F.: Amenable groups and cellular automata. Ann. Inst. Fourier (Grenoble) **49**, 673–685 (1999)
4. Gromov, M.: Endomorphisms of symbolic algebraic varieties. J. Eur. Math. Soc. (JEMS) **1**, 109–197 (1999)
5. Moore, E.F.: Machine models of self-reproduction. In: Proceedings of Symposia in Applied Mathematics, vol. 14, pp. 17–34. American Mathematical Society, Providence (1963)
6. Myhill, J.: The converse of Moore's Garden-of-Eden theorem. Proc. Amer. Math. Soc. **14**, 685–686 (1963)
7. Schmidt, K.: Dynamical Systems of Algebraic Origin, Progress in Mathematics, vol. 128. Birkhäuser Verlag, Basel (1995)

Contents

On the Effects of Firing Memory
in the Dynamics of Conjunctive Networks

Eric Goles[1], Pedro Montealegre[1], and Martín Ríos-Wilson[2(✉)]

[1] Facultad de Ingeniería y Ciencias, Universidad Adolfo Ibáñez, Santiago, Chile
{eric.chacc,p.montealegre}@uai.cl
[2] Departamento de Ingeniería Matemática,
Facultad de Ciencias Físicas y Matemáticas, Universidad de Chile, Santiago, Chile
mrios@dim.uchile.cl

Abstract. Boolean networks are one of the most studied discrete models in the context of the study of gene expression. In order to define the dynamics associated to a Boolean network, there are several *update schemes* that range from parallel or *synchronous* to *asynchronous*. However, studying each possible dynamics defined by different update schemes might not be efficient. In this context, considering some type of temporal delay in the dynamics of Boolean networks emerges as an alternative approach. In this paper, we focus in studying the effect of a particular type of delay called *firing memory* in the dynamics of Boolean networks. Particularly, we focus in symmetric (non-directed) conjunctive networks and we show that there exist examples that exhibit attractors of non-polynomial period. In addition, we study the prediction problem consisting in determinate if some vertex will eventually change its state, given an initial condition. We prove that this problem is **PSPACE**-complete.

Keywords: Boolean network · Firing memory ·
Conjunctive networks · Prediction problem · PSPACE

1 Introduction

Boolean networks are one of the simplest and most studied discrete models in the context of the study of gene expression [24,25,34]. A boolean network is defined by a boolean map that is usually represented as graph, called interaction graph, where the vertices or nodes represent genes and the edges represent regulatory interactions. A gene in the network can be active or inactive and that is represented by a node in *state* 1 or 0 respectively. This model was first introduced by Kauffman in the end of the 60's [25] and it was thought as a generalization of the

Electronic supplementary material The online version of this chapter (https://doi.org/10.1007/978-3-030-20981-0_1) contains supplementary material, which is available to authorized users.

McCulloch and Pitts neural network model [28]. The seminal papers by Kauffman and Thomas focused in studying the dynamical properties of random generated networks [23,25,26] as well as studying the structure involved in the regulatory circuitry [34,35]. A boolean network naturally defines a discrete dynamical system by updating all the nodes of the network simultaneously, i.e. the consecutive states of the dynamics are given by iterations of the original boolean map. This update scheme is called *parallel* or *synchronous*. As the number of possible states is finite (it is given by the number of possible tuples with values 0 and 1 which is 2^n) every initial state eventually exhibit periodic dynamics. We call the set of states that define these periodic dynamics an *attractor*. If the attractor is one single state we call it a *fixed point* and otherwise we call it a *limit cycle*. Though this model is fairly simple to study, it fails to reproduce gene expression data in a realistic way, mainly because of the synchronous update scheme. One straightforward approach to improve the model is adding asynchronicity to the dynamics by considering different update schemes [2,6,13,33]. Since some biologists agreed that some synchronicity is not completely unrealistic [5,37] updates schemes usually range from synchronous to sequential update schemes in which every node is updated according to a given partial order. A notable example of update schemes that are somewhere between the latter categories are *block sequential* update schemes. In these update schemes, a partition in the node set is defined and nodes inside each set of the partition are updated in parallel while sets in the partition are updated sequentially. However, in order to define one of the latter update schemes, a partition and an partial order need to be chosen. These requirements introduce, in the biological networks modelling framework, several ways to model the dynamics of a fixed object of study. Although it is relevant and interesting from a mathematical or computational point of view to study the dynamics generated by every possible update scheme in the latter context, this exercise might turn to be rather impractical.

An alternative approach to allow adding asynchronicity to the dynamics of a boolean network is based in the concept of delay that is generally defined as an internal clock, that could be independent from the original dynamics of the system, and that dictates its dynamical behaviour during a fixed time interval. This latter concept was first introduced by Thomas in [35,36] and then studied in different frameworks such as in [1,4,7,31,32]. Particularly, here we are interested in specific type of delay called *firing memory*. It was based in the concept of memory and it was first introduced by Graudenzi and Serra under the name of *gene protein Boolean networks* [16–18] and they defined this delay inspired in the concept of decay of proteins. In [9], Goles et al. introduced some modifications of the original model and presented it under the name of *Boolean networks with firing memory*. A question that naturally arise in this context is what are the effects of firing memory in the dynamics of boolean networks. According to [9], one of the first observations stated in the seminal papers, that was deduced through the analysis of numerical simulations, is that the higher the maximum time decay value (delay) the less the network admits asymptotic degrees of freedom. In order to survey this observation from a theoretical point of view, a straightforward methodology is to study a specific class of boolean networks preferably the one where the dynamics have been characterized. In fact, we are interested in the effect of *firing memory* in the dynamics of threshold networks. In these networks, the state of every node evolves accordingly to a *threshold function* that depends on the state of certain variables represented as the neighbors of the node in underlying interaction graph. In [15] Goles et al. characterized the dynamics of the latter network (without delay) and particularly they showed that attractors can only be limit cycles with period 2 or fixed points. One of the simplest type of threshold networks are the *disjunctive* and *conjunctive* boolean networks in which the state of every node depends on an OR or an AND function of its neighbors respectively. In [9], Goles et al. proved that disjunctive networks with firing memory only admit homogeneous fixed points as attractors. However, the effect of these type of delay in the dynamics of conjunctive networks have not been described until now. Perhaps surprisingly, conjunctive networks with firing memory can exhibit extremely different behaviour compared to the regular ones.

In the latter context, an interesting question is if firing memory is able to induce in the original boolean network dynamics the capability of simulating other computation models such as boolean circuits, Turing machines, etc. This line of research led us to consider a natural problem that arise in the study of boolean network dynamics: the *prediction problem*. This problem is defined in the following way: given an initial condition and an update scheme(in this case parallel scheme with firing memory), to predict the future states. To solve that problem, several strategies can be proposed from directly simulating the network to more elaborated strategies based on the topological or algebraical properties of the network. A measure of the efficiency of an strategy is given by the computational complexity of the problem. Prediction problems have been broadly studied in threshold networks [10,11] and particularly, in disjunctive (conjunctive) networks, it is known that the problem is in the **P** class.

In this paper, we focus in studying the dynamics of conjunctive networks with firing memory and we prove that, contrary to what might be assumed on previous results for disjunctive networks, conjunctive networks with firing memory admit attractors of non polynomial period. Then, we study the prediction problem and we prove that it is **PSPACE**-complete. We achieve this by showing that conjunctive networks with firing memory are capable of simulating iterated boolean circuits. As a direct corollary of this result, we conclude that the latter boolean networks with firing memory are universal, in the sense that they are able to simulate an arbitrary given boolean function.

2   Contributions and Structure of the Paper

In this paper we show that, contrarily to what one may think, conjunctive networks with firing memory exhibit an extremely complex dynamical behavior. More precisely, we show that 2-AND-PREDICTION is **PSPACE**-complete as a consequence of the capability of this rule to simulate iterated monotone boolean circuits. As a corollary of the latter result, we show that conjunctive boolean networks with firing memory are a universal model in the sense that they are capable of simulating every boolean network automata.

The rest of the paper is organized as follows. In Sect. 3 we give the main formal definitions and previous results. In Sect. 4 we show the gadgets that play an essential role in the proof of our main results and we use them to exhibit a conjunctive network with firing memory that admits attractors of non polynomial period. In Sect. 5 we study the computational complexity of the 2-AND-PREDICTION problem and we give a complete proof of the main result in Theorem 3.

3   Preliminaries

A boolean network is a map $F : \{0,1\}^n \to \{0,1\}^n$. Associated to this function, we define its interaction graph $\mathcal{G}(f) = (V, E)$ by $V = \{1, \ldots, n\}$ and $ij \in E \iff$ F_j depends on the variable x_i. F defines a dynamical system $(X = \{0,1\}^n, F)$ in which the elements $x \in X$ are called *states* or *configurations*. and the transitions are given by the iterations of the map F, i.e, for every state $x \in X$ we define its next state by $x(1) = F(x)$ and in general we have that $x(t + 1) = F(x(t))$ for every $t \in \mathbb{N}$ $(x(0) = x)$. This type of dynamics is often called *parallel* or *synchronous* update scheme. In the next sections we will assume that boolean networks dynamics will be defined in this way. Given an initial condition $x \in X$, we call its associated *trajectory* to infinite sequence $T(x) = (x(0) = x, x(1), \ldots)$. As the number of possible states is finite (2^n), every trajectory is eventually periodic, i.e., there exists $p \geq 0$ such that $x(t + p) = x(t)$. We say that a trajectory reaches a *limit cycle* with period p if the last property hold for that trajectory and p is the minimum time in which the property is satisfied. A set of configurations in a limit cycle with period p is called an *attractor* with period p. Particularly, when $p = 1$ we say that the attractor is a *fixed point*.

We are interested in some specific type of boolean networks called *threshold* networks. A *threshold* network is a boolean network in which given a matrix $A = (a_{ij})$ with integer entries and an integer vector $\Theta = (\theta_i)_i$ the function F is defined by

$$F(x)_i = \begin{cases} 1 & \text{if } \sum_{j=1}^{n} a_{ij} x_j - \theta_i \geq 0 \\ 0 & \text{otherwise} \end{cases}$$

One particular class of threshold networks are disjunctive and conjunctive networks. Disjunctive networks are defined by threshold 1 in every coordinate function F_i, in other words, are defined by an OR of certain variables i.e., $F(x)_i = F_i(x_{j_1}, \ldots, x_{j_k}) = \bigvee_{i=1}^{k} x_{j_i}$. On the other hand, a conjunctive network is a boolean network F such that every local rule F_i is given by an AND function of certain variables, i.e., $F(x)_i = F_i(x_{j_1}, \ldots, x_{j_k}) = \bigwedge_{i=1}^{k} x_{j_i}$. In this case, we have that $\theta_i = \delta_i$ for every i where δ_i is the number of neighbors of i in the interaction graph associated to F. These networks have been broadly studied in different frameworks [3, 8, 12, 22] mainly because their simplicity and their relevance in applications in modelling gene regulatory networks in which conjunctive functions describe common regulatory interactions [19, 30].

As we referred in the introduction, the concept of delay in boolean networks has emerged as an alternative approach to introduce asynchronicity. In particular, we are interested in studying the effects of a type of delay called *firing memory*. We consider a boolean network F and states $Y = \prod_{i=1}^{n} \{0, 1\} \times \{1, \ldots, dt_i\}$, $dt_i \geq 1$ for all i. Given $y(0) \in X$, $y(0)_i = (x(0)_i, \Delta(0)_i)$, we define the following dynamics:

$$x(t+1)_i = \begin{cases} 1 & \Delta(t+1)_i \geq 1, \\ F_i(x(t)) & \Delta(t+1)_i = 0. \end{cases}$$

$$\Delta(t+1)_i = \begin{cases} dt_i & F_i(x(t)) = 1, \\ \Delta(t)_i - 1 & F_i(x(t)) = 0 \wedge \Delta(t)_i \geq 1, \\ 0 & F_i(x(t)) = 0 \wedge \Delta(t)_i = 0. \end{cases}$$

This local rule $(x(t)_i, \Delta(t)_i) \to (x(t+1)_i, \Delta(t+1)_i)$ defines a global transition function $F^{dt} : Y \to Y$ that we call boolean network with *firing memory*.

Remark 1. *One useful notation introduced in [9], is considering the states as the single delay value instead of a tuple. For example, the state $(1, 2)$ that means state 1 and delay 2 is represented exclusively by 2. In the next sections, we will be using this notation.*

Remark 2. *We are implicitly assuming that a node i can not have simultaneously state $x_i = 0$ and delay $\Delta_i \neq 0$. In particular, if $y = (x, \delta) \in Y$ is an initial condition and $\delta_i = 0$ for some $i \in V$ then we must have $x_i = 0$.*

A natural question regarding the effects of firing memory in the dynamics of conjunctive networks is if this type of delay is able to give simulation capabilities to the network in the sense of allowing it to simulate other boolean networks of a different class or other computation models. In this context, we are interested in studying *prediction problems*. A well studied topic in the context of the dynamics of boolean networks is to make predictions about the attractor associated to an specific trajectory defined by a initial condition x. There exists a very simple solution to this problem that is to simulate the network dynamics until the initial state reaches a limit cycle. However, a question that naturally arise is if there exists a more efficient solution, considering the fact that using the last strategy may take as many steps as there are possible states. These more efficient solutions would be based on algorithmic or algebraic properties of the network. If $x \in \{0, 1\}^n$ we introduce the complement of x denoted by $\bar{x}$ and defined by: $x_i = 1$ implies $\bar{x}_i = 0$ and $x_i = 0$ implies $\bar{x}_i = 1$. Given a maximum delay vector dt we define the following decision problem:

dt-AND-PREDICTION:
Given a conjunctive network with firing memory F with maximun delay vector dt, $i \in \{1, \ldots, n\}$ and a configuration $y = (x, \delta) \in Y$, does there exist $z = (w, \tau) \in T(y)$ such that $w_i = \bar{x}_i$?

We remark that, because we are working with the AND rule, if some configuration (x, δ) satisfies that $x_i = 0$ (that means that its delay value δ_i is considered to be 0) and $F(x)_i = 0$ then $x(t)_i = 0$ for all $t \geq 2$ (and so $\delta(t)_i = 0$), so the dt-AND-PREDICTION problem can be solved simulating one step of the local rule. Because of the latter observation, we can assume that $x_i = 1$. In addition, when we consider the maximum delay vector as uniform, i.e. the same maximum delay $dt_i = \tau$ in every i, we will refer to dt-AND-PREDICTION as τ-AND-PREDICTION. We are interested in studying the computational complexity of the previous problem. This concept is roughly defined as the amount of resources that are needed to find a solution, given as an expression of the input size. Classical theory defines the following main classes of complexity: **P** is the class of problems solvable by a deterministic Turing machine in polynomial time and **PSPACE** is the class of problems solvable by a deterministic machine that uses polynomial space. Additionally, it is known that $\mathbf{P} \subset \mathbf{PSPACE}$. It is conjectured that these inclusions are strict, so there are problems in **PSPACE** that do not belong to **P**. The problems in **PSPACE** that are the most likely to not belong to **P** are the **PSPACE**-complete problems, which analogously **NP**-complete problems, are the ones such that any other problem in **PSPACE** can be reduced to them in polynomial time.

One very well known type of **PSPACE**-complete problem is related to the iterative evaluation of boolean circuits. A boolean circuit is a directed acyclic graph C that have three types of vertices: the ones with in-degree 0 called *inputs*, the ones with out-degree 0 called *outputs* and the rest of the vertices that have in and out neighbors called logical gates. These nodes are labelled by $\wedge, \vee, \neg$. A boolean circuit simulates a boolean function in the obvious way, and because of that, usually a circuit with n inputs and m outputs is denoted by $C : \{0,1\}^n \to \{0,1\}^m$. A circuit C is monotone if there are no gates labelled by $\neg$. For each gate of a circuit, its *layer* is the length of the shortest path from an input to the gate. A monotone circuit is alternating if for any path from an input to an output the gates on the path alternate between OR and AND gates. In addition, the inputs are connected to OR gates exclusively and outputs are OR gates. We define the following decision problem: Given a (monotone) boolean circuit $C : \{0,1\}^n \to \{0,1\}^n$, an input $x \in \{0,1\}^n$, and $i \in \{0, \ldots, n\}$ whether there exists a time $t \geq 1$ such that $C^t(x)_i = 1$. We call this problem ITER-CIRCUIT-PREDICTION (respectively ITER-MON-CIRCUIT-PREDICTION).

Proposition 1. ([11]). ITER-MON-CIRCUIT-PREDICTION *is **PSPACE**-complete even when restricted to alternating circuits of degree 4.*

3.1   Previous Results

Threshold networks were vastly studied in [14,15,21,29] and Goles et al. [15] showed using a technique based on monotone energy operator that the *synchronous* dynamics associated to these networks admits only attractors of bounded period (moreover, there are only attractors with period 2 and fixed

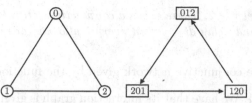

(a) Conjunctive network with firing memory and delay $dt_i = 2$ in every node that admits a limit cycle with period 3. On the left hand side we show the interaction graph of the network and on the right hand side there is the transitions graph of the cycle.

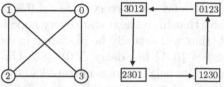

(b) Conjunctive network with firing memory and delay $dt_i = 3$ in every node that admits a limit cycle with period 4. On the left hand side we show the interaction graph of the network and on the right hand side there is the transitions graph of the cycle.

Fig. 1. Attractors with period $p = \tau + 1$ for $\tau = 2$ and $\tau = 3$

points) when the associated weight matrix is symmetric, i.e. when the underlying interaction graph is non-directed which is the case we are most interested on in this paper.

A wide studied subclass of threshold networks are the conjunctive or disjunctive networks. These systems have a very important role in modelling of biological systems because mainly because of its simplicity and their straightforward way to describe common interactions between different variables. The dynamics of these type of networks was studied under different update schemes in [12]. Our approach here is to continue the studies presented in the seminal paper of the *firing memory* [9] in which dynamics of disjunctive networks with firing memory were characterized. In the latter paper, Goles et al. showed that disjuntive networks with firing memory and delay $dt_i \geq 2$ in at least one coordinate i admit only homogeneous fixed points when the network is defined over a strongly connected directed graph.

4 Conjunctive Networks with Firing Memory Admit Non-polynomial Cycles

One surprising observation about the effect of firing memory in the dynamics of conjunctive networks is that it allows the dynamics to have attractors with period $p \geq 3$. We recall that these type of dynamics(without delay) have bounded cycles of maximum period $p = 2$ [15]. A general method to construct dynamics with a given period $p \geq 3$ is considering a conjunctive network defined by an interaction graph given by a complete graph K_{p+1} and a firing memory $dt_i = p$ in every vertex.

Proposition 2. *Let $\tau \geq 2$. There exists a conjunctive network with firing memory of size $\tau + 1$ and delay $dt_i = \tau$ in every i allowing attractors with period $p = \tau + 1$.*

Proof. Consider the conjunctive network given by the function $F(x)_i = \bigwedge_{j \neq i} x_j$, where $i \in \{0, \ldots, \tau\}$. We have that its interaction graph is given by the complete graph with $\tau + 1$ vertices $K_{\tau+1}$ (see Fig. 1). We are going to exhibit an attractor X with period $p = \tau + 1$. Let $x_0 = (0123 \ldots \tau)$ be the initial condition. Observe that every $i \in V$ has initial delay i and every node $i \neq 0$ is in state 1. Because of how we defined F we have that in the next state every node $i \neq 0$ will be set to 0. However, the only node that will actually be in state 0 in its next state is $i = 1$, because every node $i \in V \setminus \{0, 1\}$ has delay $\Delta(0)_i \geq 2$. On the other hand, we have that for $i = 0$ every of its neighbours is in state 1 in the initial condition, so it will be updated as τ in the next iteration. Thus, $x_1 = (\tau 0123 \ldots \tau - 1)$. Now, we have that every node except $i = 1$ is in state 1 and every node $i \in i \in V \setminus \{1, 2\}$ has delay $\Delta_i(1) \geq 2$ so using the same argument we used for deducing x_1 we have that $x_2 = (\tau - 1\tau 0123 \ldots, \tau - 2)$. Iterating this process $\tau + 1$ times we verify we have a cycle with period $p = \tau + 1$:

$$X = \begin{cases} x_0 = (0123 \ldots \tau - 1\tau) \\ x_1 = (\tau 0123 \ldots \tau - 1) \\ x_2 = (\tau - 1\tau 0123 \ldots, \tau - 2) \\ \quad \vdots \\ x_\tau = (123 \ldots \tau - 1\tau 0) \\ x_0 = (0123 \ldots \tau - 1\tau) \end{cases}$$

As a consequence of the last proposition, we have a stronger result on the period of the attractors when we consider conjunctive networks with firing memory with different maximum delay values.

Theorem 1. *There exists a connected conjunctive network of size n with firing memory (and not necessarily the same values for maximum delay) which admits attractors with period $2^{\Omega(\sqrt{n \log n})}$.*

Proof. Let us consider a fixed natural number $m \geq 3$ and a collection of prime numbers $p_i, \ldots, p_l$, such that $2 \leq p_i \leq m$ where $\ell = \pi(m)$ and $\pi(m)$ denotes the number of primes not exceeding m. For each $i \in \{1, \ldots, \ell\}$ we consider a conjunctive network with interaction graph given by a complete graph K_{p_i} as we did in the proof of the last proposition. We consider a graph G defined as the connected union of the previous complete graphs in the following way: we consider $V = \bigcup_{i=1}^{\ell} V(K_{p_i}) \cup \bigcup_{i=1}^{\ell-1} \{r_i\}$ and for every $i \in \{1, \ldots, \ell\}$ we choose an arbitrary vertex $s_i \in K_{p_i}$ and we consider $E = \bigcup_{i=1}^{\ell} E(K_{p_i}) \cup \bigcup_{i=1}^{\ell-1} \{s_i r_i\} \cup \bigcup_{i=1}^{\ell-1} \{s_{i+1} r_i\}$. In other words, we consider the union of the previous complete graphs and we connect each other by a unique 2-path.

Let us define the label function $\varphi : V \to \{1, \ldots, \ell\}$ given by $\varphi(u) = i$ if $u \in V(K_{p_i})$. Let $F^{p_1}, \ldots, F^{p_l}$ be the conjunctive networks with firing memory associated to each complete graph K_{p_i}. We recall that each of these functions has a maximum delay value of $p_i - 1$ in every node. We define the following conjunctive network with firing memory $F : \{0,1\}^{|V|} \to \{0,1\}^{|V|}$ given by $F(x)_i = F^{p_{\varphi(i)}}(x)_i$ for every $i \in V \setminus \{s_j | j \in \{1, \ldots, \ell\}\}$, $F(x)_{s_i} = F^{p_{\varphi(s_i)}}(x)_{s_i} \wedge x_{r_i} \wedge x_{r_{i+1}}$ for $i \in \{2, \ldots, \ell-1\}$, $F(x)_{s_\ell} = F^{p_{\varphi(s_\ell)}}(x)_{s_\ell} \wedge x_{r_{\ell-1}}$, $F(x)_{s_1} = F^{p_{\varphi(s_1)}}(x)_{s_1} \wedge x_{r_1}$ and $F(x)_{r_i} = x_{s_i} \wedge x_{s_{i+1}}$ for $i \in \{1, \ldots, \ell-1\}$. It is not difficult to see that the interaction graph of F is G. We note that functions $F(x)_{r_i}$ are regular conjunctive functions, in other words, delay values for this nodes are $dt_i = 1$.

Finally, let us define the initial condition $x \in \{0,1\}^{|V|}$ in the following way: for the vertices in $V(K_{p_i})$ we assign the initial condition $(0123, \ldots, p_i)$ and $x_{r_i} = 1$ for all $i \in 1, \ldots, \ell - 1$. Note that we cannot have two nodes connected and both having initial state 0 because of the nodes in different components are connected through the vertices r_i. Thus, the global dynamics converge to the fixed point 0. It is not difficult to see that starting from x every node in K_{p_i} is in a cycle with period p_i. Thus, we have that if T is the global period of the network then:

$$T \geq \prod_{i=1}^{\pi(m)} p_i.$$

And also, we have that:

$$|V| = \sum_{i=1}^{\pi(m)} (p_i + 1) - 1. \tag{1}$$

Additionally, if we define $\theta(m) = \sum_{i=1}^{\pi(m)} \log(p_i)$, we have that:

$$T \geq \exp(\theta(m)). \tag{2}$$

Based in 1, and 2, we are going to apply a technique used in [27] to deduce that T is not polynomial. This is based a result stated in [20]:

$$\pi(m) = \Theta\left(\frac{m}{\log(m)}\right), \tag{3}$$

$$\theta(m) = \Theta\left(\pi \log(m)\right). \tag{4}$$

We observe that from 3 it can be deduced that $m = O(\pi(m)log(m))$ and then $|V| = O(\pi(m)^2 \log(m))$. On the other hand, using this last observation we deduce that $\log |V| = O(\log(m))$. And finally, $\sqrt{|V| \log(|V|)} = O(\pi(m) \log(m))$ which is equivalent to say that $\pi(m) \log(m) = \Omega\left(\sqrt{|V| \log(|V|)}\right)$ and thus

$$T \geq \exp(\Omega\left(\sqrt{|V| \log(|V|)}\right)).$$

We conclude that F has attractors with non polynomial period.

One natural question that arises in the context of the last theorem is if we can say something about the period of the attractors in the case when we restrict a conjunctive network with firing memory to have the same delay τ in all its coordinates. Would there exist a network of this class whose dynamics allows attractors with non-polynomial period? The answer is yes, but in order to exhibit it, we need to prove a proposition that is analogous to Proposition 2.

Proposition 3. *Let $\tau \geq 2$. For every integer $k \geq 2$, there exists a conjunctive network with firing memory of size $k\tau^2(\tau + 1)$ and maximum delay $dt_i = \tau$ in every node i which admits attractors with period $k(\tau + 1)$.*

Proof. Let $k \geq 1$ be an integer. Let us define $C = K_{\tau+1}$ as complete graph with $\tau + 1$ vertices. We recall that these gadget defines a conjunctive network with firing memory which allows cycles of length $\tau + 1$ when we have that the maximum delay of every node is $dt_i = \tau$. We are going to call this structure a clock. We are going to define k copies of a certain gadget that we will call block. Let $j \in \{0, \dots, k\}$ we define the j-block B_j as a $\tau + 1$-path such that every node has exactly $\tau - 1$ neighbours and each one of these is in a different clock beside its neighbour in the path as it is shown in Fig. 2. As every block contains an induced path then we can number the vertices in the path defining an initial and a terminal vertex. We write $C(B_j)_{r,l}$ for $r = 1, 2, 3, \dots, \tau$ and $l = 1, 2, 3, \dots, \tau - 1$ to denote the clocks of the j-block that are associated to the r-th vertex of the path. Besides, as every node in the block has $\tau - 1$ neighbours(each one in a different clock), given a node r in a block B_j we will call its neighbour in the clock $C(B_j)_{r,l}$ an *anchor* of r and we will denote it by $C(B_j)_{r,l}a$ or simply by a when the context is clear. We remark that the only vertex in a clock that is connected to exactly one node in the path is its anchor. The other nodes are only connected to vertices in the clock. Finally we consider the graph G as the connected union of k-blocks defined connecting the terminal vertex of the j-th block to an initial vertex in $(j + 1)$-th block and the terminal vertex of k-block to the initial vertex of the 1st-block. With this construction G is a $k(\tau+1)$-cycle in which every node is connected to a clock. Using the structure of G, we define a global rule F^τ as a conjunctive network with firing memory and maximum delay values τ which have as underlying interaction graph G.

Now, we define the dynamics of an attractor with period $k(\tau + 1)$ using F^τ. We define the initial condition $x \in \{0, 1\}^{|V|}$ by setting the same initial condition in each block except the first one. For first block we have the state $0123 \cdots \tau$ for the nodes in the path and for $j = 2, 3, \cdots, k$ we have the initial states $\tau123 \cdots \tau$. For the clocks, we have that in each of these there is a unique anchor connected to one node in the path. We have by Proposition 2, that if we ignore this connection, every clock dynamics is associated to an attractor with period $\tau + 1$. In fact, we will show that choosing a particular initial condition, every clock will exhibit this dynamics. We are going to write only the state of the anchor a in every clock and the other states in this subgraph are assumed to be in the initial states so the clock defines an attractor with period $\tau + 1$. Assuming this notation, the initial state for the clocks are:

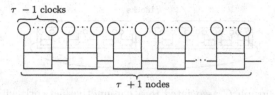

Fig. 2. Structure of the j-th block used in Proposition 2 to define a conjunctive network with firing memory and maximum delay values $dt_i = \tau$ for every node i that admits attractors with period $k(\tau + 1)$. Every circle in the figure represents clock $C(B_j)_{r,l}$ associated to a node r represented by a square. This gadget has $\tau + 1$ nodes and every node has $\tau - 1$ clocks.

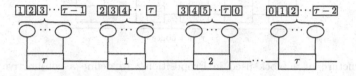

Fig. 3. Initial condition for the j-th block, $j \geq 2$ used in Proposition 2 to define an attractor with period $k(\tau + 1)$. For the first block we just define the state of the first node to 0 instead of τ.

$$\underbrace{123\ldots(\tau-1)}_{\text{Node 0}}\ \underbrace{234\ldots\tau}_{\text{Node 1}}\ \underbrace{345\ldots\tau0}_{\text{Node 2}}\ldots\ \underbrace{0123\ldots(\tau-2)}_{\text{Node }\tau}.$$

A summary of the initial condition x is shown in Fig. 3.

Assuming this initial condition, we are going to show that the global period of the network is $k(\tau + 1)$. In order to that, it suffices to show that for every node i in the network such that $x_i \neq 0$, its next state will be given by the state of its neighbour that has state $x_i - 1$ and if $x_i = 0$ then its next state will be τ. In fact, if we have this property then, the next state of the first block will be $\tau 0123\ldots\tau - 1$ and if we continue iterating we will have that in $\tau + 1$ steps the second block will have the state $0123\cdots\tau$ and so in $k(\tau + 1)$ steps the network will return to the initial condition.

Let's see that our claim is true. Take the block B_j for some $j \geq 2$. We have that every node in the path, except for the first and second node, has some anchor a in state 0 as it is shown in Fig. 4. Then, we will have that third to last nodes in the path will decrease their delay states by one and the first and second node will restore its delay to the maximum value because they have no neighbours in state 0. Thus, the next state for the nodes in the path will be $\tau\tau123\ldots\tau - 1$. Besides, as anchors in clocks have no neighbours in the path with state 0 and other vertices in the clock are not connected to the path, clocks will be updated in the following way:

$$\underbrace{012\ldots(\tau-2)}_{\text{Node 0}}\ \underbrace{123\ldots\tau-1}_{\text{Node 1}}\ \underbrace{234\ldots\tau-1\tau}_{\text{Node 3}}\ldots\ \underbrace{\tau012\ldots(\tau-2)}_{\text{Node }\tau}.$$

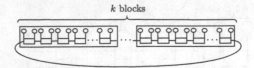

Fig. 4. Interaction graph G associated to a conjunctive network with firing memory and maximum delay $dt_i = \tau$ for all $i \in V(G)$, defined in Proposition 2 that admits attractors with length $k(\tau + 1)$.

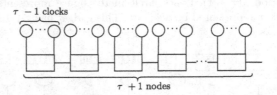

Fig. 5. Structure of a block used in Proposition 2 to define a conjunctive network with firing memory and maximum delay values $dt_i = \tau$ for every node i that admits attractors with period $k(\tau + 1)$. Every circle in the figure represents clock associated to a node r represented by a square. This gadget has $\tau + 1$ nodes and every node has $\tau - 1$ clocks.

We note that now the first node in the path (that we call node 0) and its associated anchors have the same state than the last node in the path had the previous step, and the same hold for the other nodes in the block and in the clocks (they all assume the state of its left neighbour if they are in the path and every node in the clock decreases its delay by 1, except for the one with initial state 0 that restores its maximum delay τ). In the other hand, if we study the dynamics of the first block, we have the same behaviour than before except that because of node 0 is now initially setted to 0, then node 1 has a neighbour that is initially setted to 0 and so in the next step, every node in the path is going descend to the previous state except for the first one that will be setted to τ, In addition, all the clocks associated to the first block have the same behaviour than before except for the ones associated to the first node in the path. Nevertheless, because every anchor associated to the latter is in a state different than 0 and also because they all already have a neighbour in 0 inside the clock, they have the same transition than in the latter analysis. Thus, the next state of the nodes in the path will be $\tau 0123 \ldots \tau - 1$ as desired and every one of these has now anchors in the same state than the previous state associated to its left neighbour. We conclude that F admits attractors with length $k(\tau + 1)$.

An immediate consequence of Proposition 3 is that we can exhibit a conjunctive network with firing memory (that have the same maximum delay values in every coordinate) which admits attractors with non polynomial period. This is possible because we can replicate, in this context, the same strategy we used to prove Theorem 1.

Theorem 2. *Let $\tau \geq 2$. There exists a conjunctive network of size n with firing memory and maximum delay $dt_i = \tau$ in every node i which admits attractors with period $2^{\Omega(\sqrt{n \log n})}$.*

Proof. Let $m \geq 2$ and a collection of prime numbers $p_1, \ldots, p_l$ where $l = \pi(m)$ as in Theorem 1. As a consequence of Proposition 3 there exist functions F_{p_i} and associated graphs C_{p_i}, $i = 1, 2, \cdots, l$ which admit attractors with period $p_i(\tau + 1)$. As same as we did to prove Theorem 1 we define a global function F with delay τ in every coordinate which associated interaction graph $G = (V, E)$ is given by the connected union of the graphs C_{p_i}. In this case, we connect every of these components by adding an edge between a node labelled by a associated to the first vertex in the path of the first block of C_{p_i} to another a labelled vertex associated to the second vertex in the path of the first block of $C_{p_{i+1}}$ as it is shown in Fig. 6. Initial condition x is defined as every node in $C_{p_{i+1}}$ is in an attractor with period $(\tau + 1)p_i$ by using the same initial condition given in the proof of Proposition 3. We remark that there no connected vertices with state 0 in every iteration because of how we defined the connection between components. Again, as same as in Theorem 1 we have that the global period of the network T satisfy that:

$$T \geq (\tau + 1) \prod_{i=1}^{\pi(m)} p_i.$$

And also, we have that:

$$|V| = (1 + (\tau - 1)(\tau + 1))(\tau + 1) \sum_{i=1}^{\pi(m)} p_i. \tag{5}$$

It is not difficult to see that applying the same technique that we used in the proof of Theorem 1 we can conclude:

$$T \geq \exp(\Omega(\sqrt{|V| \log(|V|)})).$$

Thus, F has attractors with non polynomial period.

5   2-And-Prediction is PSPACE-complete.

The fact that there exist conjunctive networks with firing memory that admit attractors with non polynomial period together with the structure of the gadgets we described, strongly suggest that there are conjunctive networks that are able to simulate boolean circuits. The next results confirm this insight establishing that conjunctive networks with firing memory are able to simulate iterated boolean circuits. We remark that previous results on the attractors period hold for an arbitrary value for the maximum delay that is defined for all the nodes in the given networks. So, we address the capability of conjunctive networks

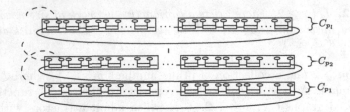

Fig. 6. Interaction graph G associated to a conjunctive network with firing memory and maximum delay values $dt_i = \tau$ for every node $i \in V(G)$ that admits attractors with non polynomial period. Every component defines a local dynamics with period $(\tau + 1)p_i$. Initial condition is defined verifying that there are no connected nodes in 0. Global period of the network is given by the product of prime numbers p_i.

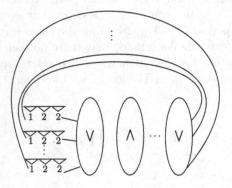

Fig. 7. Interaction graph associated to a conjunctive network with firing memory and maximum delay vector $dt_i = 2$ for every node i that simulates an iterated monotone boolean circuit C. Layers are made up by AND or OR gates exclusively, using the gadgets shown in Figs. 8 and 9 are alternately ordered.

with firing memory and maximum delay $dt_i = 2$ in every node i to simulate an arbitrary iterated boolean circuit. These results have consequences related to computational complexity and universality of conjunctive networks with firing memory.

Proposition 4. *For every monotone boolean circuit $C : \{0,1\}^n \rightarrow \{0,1\}^n$ there is a conjunctive network F with firing memory such that: i) its interaction graph G has linear size in n, ii) its maximum delay values are $dt_i = 2$ in every node $i \in V(G)$ and iii) F simulates each iteration of C in linear time.*

Proof. Let C be a monotone circuit. We assume that: (i) every input has out degree 1, (ii) every output is identified with an input, (iii) the degree of every logic gate in C is 4, and (iv) every layer contains exclusively OR or AND gates and they are ordered alternately, i.e., if the k-th layer is made up of AND gates then the $k + 1$-th is made up of OR gates (see Proposition 1). We represent this structure using the block gadget we defined for the last propositions (see Fig. 5)

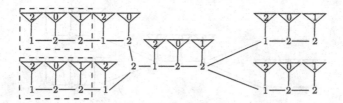

Fig. 8. Gadget of AND gates used in the graph shown in Fig. 7. Signals are transmitted and coded based on the block gadget.

with maximum delay vector $dt_i = 2$ in every node i. A scheme of the interaction graph associated to this conjunctive network with firing memory is shown in Fig. 7. Let $i_1, \ldots, i_n$ be the inputs of C and because we are considering C as an iterated circuit, we are going to identify its outputs by the same names. For every $k \in \{1, \ldots, n\}$, we define a block B_{i_k}. These blocks are made up of a path of length three in which every node is connected to a clock. We introduce the following notation: if $x \in \{0, 1\}^n$ and B is a block then $x_B = x_u x_v x_w$ where u, v and w are the labels of the vertices in the path. Given a initial configuration $x \in \{0, 1\}^n$ for the circuit C, we code it using a variable $y \in \{0, 1\}^m$, defined by mapping the blocks in the following way:

$$y_{B_{i_k}} = \begin{cases} 122 \text{ if } x_{i_k} = 0 \\ 120 \text{ if } x_{i_k} = 1 \end{cases}$$

For every logic gate AND or OR we define a gadget as the one showed in Figs. 8 and 9. Note that every logic gate is represented by a block so the last coding is well defined. We remark that as all the gadgets have a constant number of nodes and edges then, $m = O(n)$ and because of the assumptions we are doing on C we have that this coding has polynomial size in n. We will call φ the function such that $\varphi(x) = y$. Finally, as it is shown in Figs. A.1 and A.2 the information is transmitted through the blocks in a way such that in a maximum of 9 steps the nodes return to the initial condition so the circuit is cleared and the structures are available for continuing receiving and emitting signals. Then, we have given $x \in \{0, 1\}^n$ and $y = \varphi(x)$, there exists $p \geq 1$ such that $\varphi(C^t(x))_{B_{i_k}} = (F^{pt}(y))_{B_{i_k}}$ for all $t \geq 0$ and $k \in \{1, \ldots, n\}$. Thus, the conjunctive network defined using these gadgets simulates C in linear space and linear time.

As a direct consequence of the Proposition 4, we have that conjunctive networks with firing memory are universal, i.e, they can simulate every boolean network. Finally, we address the question about the computational complexity of the prediction problem 2-AND-PREDICTION. As a direct corollary of the latter proposition, we have that ITER-MON-CIRCUIT can be reduced to 2-AND-PREDICTION and thus the problem is **PSPACE**-complete.

Theorem 3. *The problem* 2-AND-PREDICTION *is* **PSPACE**-*complete.*

Proof. It is direct from Propositions 4 and 1

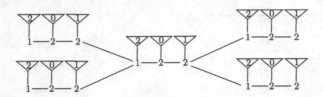

Fig. 9. Gadget of OR gates used in the graph shown in Fig. 7. Signals are transmitted and coded based on the block gadget.

6   Discussion

In this paper, we have studied the effect of specific type of delay called firing memory in the dynamics of conjunctive boolean networks. More specifically, we have addressed the prediction problem in conjunctive networks with firing memory whose maximum delay is 2 in every node. We concluded that not only these type of networks admit attractors of non polynomial period but the latter problem turned out to be **PSPACE**-complete. Deducing this result was possible because of: (i) the capability of conjunctive networks with firing memory that have the same value for maximum delay in every node, to have attractors with an arbitrary long period (Proposition 3) and (ii) the capability of transmitting information through a wire that clears itself once the information has been transmitted (construction of blocks and clocks). These two main observations about conjunctive networks with firing memory allowed us to deduce the structure of the main gadgets used for the proof of our main results. These properties are quite surprising considering that conjunctive boolean networks admit only attractors of bounded period and the prediction problem is in **P**. Moreover, previous result on the effect of firing memory in the dynamics of the dual version of these type of networks, the disjunctive networks, suggested that firing memory tend to freeze the dynamics of these networks, reducing the period of the possible attractors that the network admits. We remark the relevance of the achieved results as they show that firing memory have effects on the dynamical properties of the original network that are similar to the ones exhibited by other update schemes that are somehow between synchronous and asynchronous dynamics in other type of boolean networks, such as the effects of block sequential update scheme in majority rules. In fact, in the latter case, the prediction problem with parallel update is **P**-complete while it is **PSPACE**-complete when we consider a block-sequential update scheme. It might be possible to deduce from the latter observation that firing memory allows to add asynchronicity to the dynamics of a boolean network in a less arbitrary way compared to block sequential update, that needs a predefined partition and an specific partial order.

An interesting topic for future work is the characterization of the dynamics of conjunctive networks with firing memory. While we have described conjunctive networks with firing memory that admit attractors with period proportional to the maximum delay values, the question about the existence of networks admitting attractors with period not necessarily a multiple of their maximum delay values remains still open. In addition, the effect of firing memory in networks defined by particular topologies such as planar graphs or two dimensional grids might be interesting to analyse. Besides, considering the fact that in the light of the results of this paper there is no clear insight about a general effect of firing memory in a simple class of boolean functions such as AND or OR (we have complex dynamics in one case and we have dynamics that admit only fixed points in the other), an interesting topic for future work could be studying prediction problems in other classes of boolean networks with firing memory that are somehow similar to conjunctive networks (other functions that are linear for example) such as XOR networks.

Acknowledgement. This work has been partially supported by: CONICYT via PAI + Convocatoria Nacional Subvención a la Incorporación en la Academia Año 2017 + PAI77170068 (P.M.), CONICYT via PFCHA/DOCTORADO NACIONAL/2018 – 21180910 + PIA AFB 170001 (M.R.W) and ECOS C16E01 (E.G and M.R.W). We would also like to thank Alejandro Maass who provided insight and expertise that greatly assisted us in the course of this research. In addition, we would like to thank the anonymous reviewers for their helpful feedback and comments, particularly in the proofs of Propositions 2 and 3.

References

1. Ahmad, J., Roux, O., Bernot, G., Comet, J.P., Richard, A.: Analysing formal models of genetic regulatory networks with delays. Int. J. Bioinform. Res. Appl. **4**(3), 240–262 (2008)
2. Aracena, J., Goles, E., Moreira, A., Salinas, L.: On the robustness of update schedules in boolean networks. Biosystems **97**(1), 1–8 (2009)
3. Aracena, J., Richard, A., Salinas, L.: Fixed points in conjunctive networks and maximal independent sets in graph contractions. J. Comput. Syst. Sci. **88**, 145–163 (2017)
4. Bernot, G., Comet, J.P., Richard, A., Guespin, J.: Application of formal methods to biological regulatory networks: extending Thomas' asynchronous logical approach with temporal logic. J. Theor. Biol. **229**(3), 339–347 (2004)
5. Boettiger, A.N., Levine, M.: Synchronous and stochastic patterns of gene activation in the Drosophila embryo. Science **325**(5939), 471–473 (2009)
6. Demongeot, J., Elena, A., Sené, S.: Robustness in regulatory networks: a multidisciplinary approach. Acta Biotheoretica **56**(1–2), 27–49 (2008)
7. Fromentin, J., Eveillard, D., Roux, O.: Hybrid modeling of biological networks: mixing temporal and qualitative biological properties. BMC Syst. Biol. **4**(1), 79 (2010)
8. Gao, Z., Chen, X., Başar, T.: Controllability of conjunctive boolean networks with application to gene regulation. IEEE Trans. Control Netw. Syst. **5**(2), 770–781 (2018)
9. Goles, E., Lobos, F., Ruz, G.A., Sené, S.: Attractor landscapes in Boolean networks with firing memory. Nat. Comput. (2019, accepted)
10. Goles, E., Montealegre, P.: Computational complexity of threshold automata networks under different updating schemes. Theor. Comput. Sci. **559**, 3–19 (2014)
11. Goles, E., Montealegre, P., Salo, V., Törmä, I.: Pspace-completeness of majority automata networks. Theor. Comput. Sci. **609**, 118–128 (2016)

12. Goles, E., Noual, M.: Disjunctive networks and update schedules. Adv. Appl. Math. **48**(5), 646–662 (2012)
13. Goles, E., Salinas, L.: Comparison between parallel and serial dynamics of boolean networks. Theor. Comput. Sci. **396**(1–3), 247–253 (2008)
14. Goles-Chacc, E.: Comportement oscillatoire d'une famille d'automates cellulaires non uniformes. Ph.D. thesis, Institut National Polytechnique de Grenoble-INPG, Université Joseph-Fourier (1980)
15. Goles-Chacc, E., Fogelman-Soulié, F., Pellegrin, D.: Decreasing energy functions as a tool for studying threshold networks. Discrete Appl. Math. **12**(3), 261–277 (1985)
16. Graudenzi, A., Serra, R.: A new model of genetic network: the Gene Protein Boolean network. In: Artificial Life and Evolutionary Computation, pp. 283–291. World Scientific (2010)
17. Graudenzi, A., Serra, R., Villani, M., Colacci, A., Kauffman, S.A.: Robustness analysis of a Boolean model of gene regulatory network with memory. J. Comput. Biol. **18**(4), 559–577 (2011)
18. Graudenzi, A., Serra, R., Villani, M., Damiani, C., Colacci, A., Kauffman, S.A.: Dynamical properties of a boolean model of gene regulatory network with memory. J. Comput. Biol. **18**(10), 1291–1303 (2011)
19. Gummow, B.M., Scheys, J.O., Cancelli, V.R., Hammer, G.D.: Reciprocal regulation of a glucocorticoid receptor-steroidogenic factor-1 transcription complex on the Dax-1 promoter by glucocorticoids and adrenocorticotropic hormone in the adrenal cortex. Mol. Endocrinol. **20**(11), 2711–2723 (2006)
20. Hardy, G.H., Wright, E.M.: An Introduction to the Theory of Numbers. Oxford University Press, Oxford (1979)
21. Hopfield, J.J.: Neural networks and physical systems with emergent collective computational abilities. Proc. Nat. Acad. Sci. **79**(8), 2554–2558 (1982)
22. Jarrah, A.S., Laubenbacher, R., Veliz-Cuba, A.: The dynamics of conjunctive and disjunctive boolean network models. Bull. Math. Biol. **72**(6), 1425–1447 (2010)
23. Kauffman, S.: Homeostasis and differentiation in random genetic control networks. Nature **224**(5215), 177 (1969)
24. Kauffman, S.: The large scale structure and dynamics of gene control circuits: an ensemble approach. J. Theor. Biol. **44**(1), 167–190 (1974)
25. Kauffman, S.A.: Metabolic stability and epigenesis in randomly constructed genetic nets. J. Theor. Biol. **22**(3), 437–467 (1969)
26. Kaufmann, S.: Gene regulation networks: a theory of their global structure and behaviour. Curr. Top. Dev. Biol. **6**, 145–182 (1971)
27. Kiwi, M.A., Ndoundam, R., Tchuente, M., Goles-Chacc, E.: No polynomial bound for the period of the parallel chip firing game on graphs. Theor. Comput. Sci. **136**(2), 527–532 (1994)
28. McCulloch, W.S., Pitts, W.: A logical calculus of the ideas immanent in nervous activity. Bull. Math. Biophys. **5**(4), 115–133 (1943)
29. Mortveit, H., Reidys, C.: An Introduction to Sequential Dynamical Systems. Springer, Boston (2008). https://doi.org/10.1007/978-0-387-49879-9
30. Nguyen, D.H., D'haeseleer, P.: Deciphering principles of transcription regulation in eukaryotic genomes. Mol. Syst. Biol. **2**(1) (2006)
31. Ren, F., Cao, J.: Asymptotic and robust stability of genetic regulatory networks with time-varying delays. Neurocomputing **71**(4–6), 834–842 (2008)
32. Ribeiro, T., Magnin, M., Inoue, K., Sakama, C.: Learning delayed influences of biological systems. Front. Bioeng. Biotechnol. **2**, 81 (2015)

33. Robert, F.: Discrete Iterations: A Metric Study. Springer, Berlin (1986). https://doi.org/10.1007/978-3-642-61607-5
34. Thomas, R.: Boolean formalization of genetic control circuits. J. Theor. Biol. **42**(3), 563–585 (1973)
35. Thomas, R.: Regulatory networks seen as asynchronous automata: a logical description. J. Theor. Biol. **153**(1), 1–23 (1991)
36. Thomas, R., Thieffry, D., Kaufman, M.: Dynamical behaviour of biological regulatory networks–I. Biological role of feedback loops and practical use of the concept of the loop-characteristic state. Bull. Math. Biol. **57**(2), 247–276 (1995)
37. Wang, Q., Huang, J., Zhang, X., Wu, B., Liu, X., Shen, Z.: The spatial association of gene expression evolves from synchrony to asynchrony and stochasticity with age. PloS One **6**(9), e24076 (2011)

Complexity-Theoretic Aspects
of Expanding Cellular Automata

Augusto Modanese[(✉)]

Institute of Theoretical Informatics (ITI), Karlsruhe Institute of Technology (KIT),
Am Fasanengarten 5, 76131 Karlsruhe, Germany
modanese@kit.edu

Abstract. The expanding cellular automata (XCA) variant of cellular automata is investigated and characterized from a complexity-theoretical standpoint. The respective polynomial-time complexity class is shown to coincide with $\leq_{tt}^{p}(\mathbf{NP})$, that is, the class of decision problems polynomial-time truth-table reducible to problems in $\mathbf{NP}$. Corollaries on select XCA variants are proven: XCAs with multiple accept and reject states are shown to be polynomial-time equivalent to the original XCA model. Meanwhile, XCAs with diverse acceptance behavior are classified in terms of $\leq_{tt}^{p}(\mathbf{NP})$ and the Turing machine polynomial-time class $\mathbf{P}$.

1 Introduction

Traditionally, cellular automata (CAs) are defined as a rigid and immutable lattice of cells; their behavior is dictated exclusively by a local transition function operating on homogeneous local configurations. This can be generalized, for instance, by mutable neighborhoods [17] or by endowing CAs with the ability to *shrink*, that is, delete their cells [18]. When shrinking, the automaton's structure and dimension are preserved by "gluing" the severed parts and reconnecting their delimiting cells as neighbors. When employed as language recognizers, shrinking CAs (SCAs) can be more efficient than standard CAs [10,18].

Other variants of CAs with dynamically reconfigurable neighborhoods have emerged throughout the years. In the case of two-dimensional CAs, there is the structurally dynamical CA (SDCA) [7], in which the connections between neighbors are created and dropped depending on the local configuration. In the one-dimensional case, further variants in this sense are considered in the work of Dubacq [5], where one finds, in particular, CAs whose neighborhoods vary over time. Dubacq also proposes the dynamically reconfigurable CA (DRCA), a CA whose cells are able to exchange their neighbors for neighbors of their neighbors. Dantchev [4] later points out a drawback in the definition of DRCAs and proposes an alternative dubbed the dynamic neighborhood CA (DNCA).

Parts of this paper have been submitted [13] in partial fulfillment of the requirements for the degree of Master of Science at the Karlsruhe Institute of Technology (KIT).

© IFIP International Federation for Information Processing 2019
Published by Springer Nature Switzerland AG 2019
A. Castillo-Ramirez and P. P. B. de Oliveira (Eds.): AUTOMATA 2019, LNCS 11525, pp. 20–34, 2019.
https://doi.org/10.1007/978-3-030-20981-0_2

By relaxing the arrangement of cells as a lattice, CAs may be generalized to graphs. Graph automata [21] are related to CAs in that each vertex in the graph can be identified as a cell; thus, graphs whose vertices have finite degrees provide a natural generalization of CAs. In [21], the authors also define a rule based on topological refinements of graphs, which may be used as a model for biological cell division. An additional example of cell division in this sense is found in the "inflating grid" described in [1].

Modeling cell division and growth, in fact, was one of the driving motivations towards the investigation of the *expanding CA* (XCA) in [14]. An XCA is, in a way, the opposite of an SCA; instead of cells vanishing, new cells can emerge between existing ones. This operation is topologically similar to the cell division of graph automata; as in the SCA model, however, it maintains the overall arrangement and connectivity of the automaton's cells as similar as possible to that of standard CAs (i.e., a bi-infinite, one-dimensional array of cells).

We mention a few aspects in which XCAs differ from the aforementioned variants. Contrary to SDCAs [7] or CAs with dynamic neighborhoods such as DRCAs [5] and DNCAs [4], XCAs enable the creation of new cells, not simply new links between existing ones. In addition, the XCA model does not focus as much on the reconfiguration of cells; in it, the neighborhoods are homogeneous and predominantly immutable. Furthermore, in contrast to the far more general graph automata [21], XCAs are still one-dimensional CAs; this ensures basic CA techniques (e.g., synchronization) function the same as they do in standard CAs.

Finally, shrinking and expanding are not mutually exclusive. Combining them yields the shrinking and expanding CA (SXCA). The polynomial-time class of SXCA language deciders was shown in [14,15] to coincide with **PSPACE**.

In [14], the polynomial-time class **XCAP** of XCA language deciders is shown to contain both **NP** and **coNP** while being contained in **PSPACE**. A precise characterization of **XCAP**, however, remained outstanding. Such was the topic of the author's master's thesis [13], the results of which are summarized in this paper. The main result is **XCAP** being equal to the class of decision problems which are polynomial-time truth-table reducible to **NP**, denoted $\leq_{tt}^{p}(\mathbf{NP})$.

The rest of this paper is organized as follows: Sect. 2 covers the fundamental definitions and results needed for the subsequent discussions. Following that, Sect. 3 recalls the main result of [14] concerning **XCAP** and presents the aforementioned characterization of **XCAP**. Section 4 covers some immediate corollaries, in particular by considering an XCA variant with multiple accept and reject states as well as two other variants with diverse acceptance conditions. Finally, Sect. 5 concludes.

2   Basic Definitions

This section recalls basic concepts and results needed for the proofs and discussions in the later sections and is broken down in two parts. The first is concerned with basic topics regarding formal languages, Turing machines, and Boolean formulas. The second part covers the definition of expanding CAs.

2.1 Formal Languages and Turing Machines

It is assumed the reader is familiar with the concepts of $\omega\omega$-words and their homomorphisms as well as deterministic and non-deterministic Turing machines (TMs and NTMs, respectively) and the fundamental classes **P**, **NP**, **coNP**, and **PSPACE**. In this paper, it is assumed all words have length at least one. The notion of a *complete* language is employed strictly in the sense of polynomial-time many-one (i.e., Karp) reductions by Turing machines.

Boolean Formulas. Let V be a language of *variables* over an alphabet Σ which, without loss of generality, is disjoint from $\{T, F, \neg, \wedge, \vee, (,)\}$. BOOL_V denotes the language of Boolean formulas over the variables of V. An *interpretation* of V is a map $I\colon V \to \{T, F\}$. Each interpretation I gives rise to an *evaluation* $E_I\colon \mathrm{BOOL}_V \to \{T, F\}$ which, given a formula $f \in \mathrm{BOOL}_V$, substitutes each variable $x \in V$ with the truth value $I(x)$ and reduces the resulting formula using standard propositional logic. A formula f is *satisfiable* if there is an interpretation I such that $E_I(f) = T$; f is a *tautology* if this holds for all interpretations.

In order to define the languages SAT of satisfiable formulas and TAUT of tautologies, a language V of variables must be agreed on. In this paper, variables are encoded as binary strings prefixed by a special symbol x, that is, $V = \{x\} \cdot \{0,1\}^+$. The language SAT contains exactly the satisfiable formulas of BOOL_V. Similarly, TAUT contains exactly the tautologies of BOOL_V. The following is a well-known result concerning SAT and TAUT [3]:

Theorem 1. SAT *is* **NP**-*complete, and* TAUT *is* **coNP**-*complete.*

Truth-Table Reductions. The theory of truth-table reductions was established in [11,12]; later it was shown the class of decision problems polynomial-time truth-table (i.e., Boolean circuit) reducible to **NP**, denoted $\leq_{tt}^p(\mathbf{NP})$, is equivalent to that of those polynomial-time Boolean formula reducible to **NP** [22]. We refer to [2] for a series of alternative characterizations of $\leq_{tt}^p(\mathbf{NP})$. As a cursory remark, we state the inclusions $\mathbf{NP} \cup \mathbf{coNP} \subseteq \leq_{tt}^p(\mathbf{NP})$ and $\leq_{tt}^p(\mathbf{NP}) \subseteq \mathbf{PSPACE}$ hold.

For the results of this paper, a formal treatment of the class $\leq_{tt}^p(\mathbf{NP})$ is not necessary; it suffices to note $\leq_{tt}^p(\mathbf{NP})$ has complete languages. In particular, we are interested in Boolean formulas with **NP** and **coNP** predicates. To this end, we employ SAT and TAUT to define membership predicates of the form $f \in_L$, where f is a Boolean formula, $L \in \{\mathrm{SAT}, \mathrm{TAUT}\}$, and "$\in_L$" is a purely syntactic construct which stands for the statement "$f \in L$".

Definition 1 ($\mathbf{SAT}^\wedge$-$\mathbf{TAUT}^\vee$). *Let* $V = \{x\}\cdot\{0,1\}^+$ *and* $V_L = \mathrm{BOOL}_V \cdot \{\in_L\}$ *for* $L \in \{\mathrm{SAT}, \mathrm{TAUT}\}$. *The language* $\mathrm{BOOL}_{\mathrm{SAT,TAUT}}^{\wedge\vee} \subseteq \mathrm{BOOL}_{V_{\mathrm{SAT}}\cup V_{\mathrm{TAUT}}}$ *is defined recursively as follows:*

1. $V_{\mathrm{SAT}}, V_{\mathrm{TAUT}} \subseteq \mathrm{BOOL}_{\mathrm{SAT,TAUT}}^{\wedge\vee}$
2. *If* $v \in V_{\mathrm{SAT}}$ *and* $f \in \mathrm{BOOL}_{\mathrm{SAT,TAUT}}^{\wedge\vee}$, *then* $\wedge(v, f) \in \mathrm{BOOL}_{\mathrm{SAT,TAUT}}^{\wedge\vee}$

3. If $v \in V_{\text{TAUT}}$ and $f \in \text{BOOL}^{\wedge \vee}_{\text{SAT,TAUT}}$, then $\vee(v, f) \in \text{BOOL}^{\wedge \vee}_{\text{SAT,TAUT}}$

The language $\text{SAT}^{\wedge}\text{-TAUT}^{\vee} \subseteq \text{BOOL}^{\wedge \vee}_{\text{SAT,TAUT}}$ contains all formulas which are true under the interpretation mapping $f \models_L$ to the truth value of the statement "$f \in L$".

The following follows from the results of Buss and Hay [2]:

Theorem 2. $\text{SAT}^{\wedge}\text{-TAUT}^{\vee}$ is $\leq^p_{tt}(\mathbf{NP})$-complete.

2.2 Cellular Automata

In this paper, we are strictly interested in one-dimensional cellular automata (CAs) with the standard neighborhood and employed as language deciders. CA deciders possess a *quiescent state* q; cells which are not in this state are said to be *active* and may not become quiescent. The input for a CA decider is provided in its initial configuration surrounded by quiescent cells. As deciders, CAs are Turing complete, and, more importantly, CAs can simulate TMs in real-time [19]. Conversely, it is known a TM can simulate a CA with time complexity t in time at most t^2. A corollary is that the CA polynomial-time class equals $\mathbf{P}$.

Expanding Cellular Automata. First considered in [14], the expanding CA (XCA) is similar to the shrinking CA (SCA) in that it is dynamically reconfigurable; instead of cells being deleted, however, in an XCA new cells emerge between existing ones. This does not alter the underlying topology, which remains one-dimensional and biinfinite.

For modeling purposes, the new cells are seen as *hidden* between the original (i.e., *visible*) ones, with one hidden cell placed between any two neighboring visible cells. These latter cells serve as the hidden cell's left and right neighbors and are referred to as its *parents*. In each CA step, a hidden cell observes the states of its parents and either assumes a non-hidden state, thus becoming visible, or remains hidden. In the former case, the cell assumes the position between its parents and becomes an ordinary cell (i.e., visible), and the parents are reconnected so as to adopt the new cell as a neighbor. Visible cells may not become hidden.

Definition 2 (XCA). *Let* $N = \{-1, 0, 1\}$ *be the standard neighborhood. An expanding CA (XCA) is a CA A with state set Q and local transition function $\delta \colon Q^N \to Q$ and which possesses a distinguished hidden state $\odot \in Q$. For any local configuration $\ell \colon N \to Q$, $\delta(\ell) = \odot$ is allowed only if $\ell(0) = \odot$.*

Let $c \colon \mathbb{Z} \to Q$ be a global configuration and $z \in \mathbb{Z}$, and let $h_c \colon \mathbb{Z} \to Q^N$ be such that $h_c(z)(-1) = c(z)$, $h_c(z)(0) = \odot$, and $h_c(z)(1) = c(z+1)$. Define $\alpha \colon Q^{\mathbb{Z}} \to Q^{\mathbb{Z}}$ as follows, where Δ is the standard CA global transition function:

$$\alpha(c)(z) = \begin{cases} \Delta(c)(\frac{z}{2}), & z \text{ even} \\ \delta(h_c(\frac{z-1}{2})), & \text{otherwise} \end{cases}$$

Finally, let Φ be the $\omega\omega$-word homomorphism induced by the mapping $Q \to Q^*$ which maps any state to itself except for $\odot$, which is mapped to ε (i.e., the empty word). Then the global transition function of an XCA is $\Delta^X = \Delta^X_\delta := \Phi \circ \alpha$.

Figure 1 illustrates an XCA A and its operation for input 001010 as an example. The local transition function δ of A is as follows:

$$\delta(q_{-1}, q_0, q_1) = \begin{cases} q_{-1} \oplus q_1, & q_{-1}, q_1 \in \{0, 1\} \\ q_0, & \text{otherwise} \end{cases}$$

(where $\oplus$ denotes the bitwise XOR operation, that is, addition modulo 2). In the initial configuration c, the hidden cells are all in the state $\odot$. Using h_c as the hidden cells' local configurations, α applies δ to each local configuration and promotes all hidden cells to ordinary (i.e., visible) ones. Finally, Φ eliminates hidden cells which conserved the state $\odot$ (as these are present only implicitly in the global configuration).

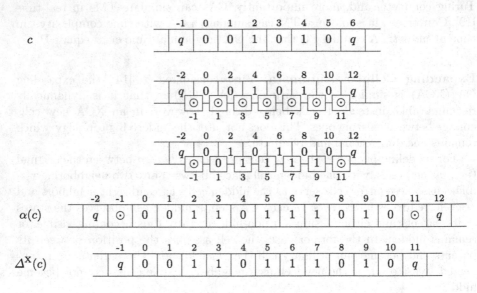

Fig. 1. Illustration of a step of the XCA A. The number next to each cell indicates its index in the respective configuration. In this particular example, we use the convention that the $\omega\omega$-word homomorphism Φ contracts deleted symbols towards index zero. Adapted from [14].

The supply of hidden cells is never depleted; if a hidden cell becomes an ordinary cell, new ones appear between it and its neighbors. Thus, the number of active cells in an XCA may increase exponentially [14]:

Lemma 1. *Let A be an XCA. For an input of size n, A has at most $(n+3)2^t - 3$ active cells after $t \in \mathbb{N}_0$ steps. This upper bound is sharp.*

We have postponed defining the acceptance behavior of an XCA until now. Usually, a CA possesses a distinguished cell, often cell 0, which dictates the automaton's accept or reject response [9]. In the case of XCAs, however, under a reasonable complexity-theoretical assumption (i.e., $\mathbf{P} \neq \leq_{tt}^{p}(\mathbf{NP})$) such an acceptance behavior results in XCAs not making full use of the efficient cell growth indicated in Lemma 1 (see Sect. 4.3). This phenomenon does not occur if the acceptance behavior is defined based on *unanimity*, that is, in order for an XCA to accept (or reject), *all* its cells must assume the accept (or reject) state simultaneously. This acceptance condition is by no means novel [6,8,16, 20]. As an aside, note all (reasonable) CA time complexity classes (including, in particular, linear- and polynomial-time) remain invariant when using this acceptance condition instead of the standard one.

Definition 3 (Acceptance behavior, time complexity). *Each XCA has a (unique) accept state a and a (unique) reject state r. An XCA A halts if all active (and non-hidden) cells are either all in state a, in which case the XCA accepts, or they are all in state r, in which case it rejects; if neither is the case, the computation continues. L(A) denotes the set of words accepted by A.*

The time complexity *of an XCA is the number of elapsed steps until it halts. An XCA decider is an XCA which halts on every input. A language L is in* **XCAP** *if there is an XCA decider A' with polynomial time complexity (in the length of its input) and such that L = L(A').*

In summary, the decision result of an XCA decider is the one indicated by the first configuration in which its active cells are either all in the accept or all in the reject state. This agrees with our aforementioned notion of a *unanimous* decision.

3   Characterizing the XCA Polynomial-Time Class

In [14], the following first result regarding the class **XCAP** is proven:

Theorem 3. $\mathbf{NP} \cup \mathbf{coNP} \subseteq \mathbf{XCAP}$.

We give a brief outline of the proof and refer the interested reader to [14] for the details. Since many-one reductions by TMs can be simulated by (X)CAs in real-time, it suffices to show **XCAP** contains **NP**- and **coNP**-complete problems. We construct XCAs for SAT and TAUT which run in polynomial time and apply Theorem 1. The two constructions are very similar; in fact, one obtains one from the other simply by swapping the accept and reject states. The following describes the XCA A for TAUT.

XCAs allow for efficient creation of new cells between existing ones. This may be used in order to efficiently create copies of the original formula and in each copy set a given variable to a possible truth value (i.e., "true" or "false"). A iterates over the input formula's variables and, at each iteration step, creates two copies of each formula: one in which the respective variable is set to "true" and one in which it is set to "false". The copies are synchronized with each other, and the process continues in parallel as separate computation branches. The synchronization ensures that, since all copies have equal length, all variables will be exhausted by each branch at the same time.

When this is the case, parallel evaluations of the resulting formulas are carried out, and their results combining using a subtlety of the accepting behavior of XCAs: Reject states are conserved while accept states yield to reject in the next step. If the result of an evaluation is "true", then all respective cells are synchronized and enter the accept state; otherwise, that is, if the result is "false", then all respective cells enter the reject state (after synchronization). As a result, if no reject states are present, then A immediately accepts; otherwise, in the next step, any existing accepting cells become rejecting cells, and A rejects.

Note the steps described above are all carried out in polynomial time since the whole process amounts to replacing variables with truth values and then evaluating the resulting formulas.

3.1 A First Characterization

This section covers the following (main) result of [13]:

Theorem 4. XCAP $= \leq_{tt}^{p}(\mathbf{NP})$.

The equality in Theorem 4 is proven by considering the two inclusions.

Proposition 1. $\leq_{tt}^{p}(\mathbf{NP}) \subseteq \mathbf{XCAP}$.

Proof. We construct an XCA A which decides $\mathrm{SAT}^{\wedge}\text{-}\mathrm{TAUT}^{\vee}$ (see Definition 1 and Theorem 2) in polynomial time. The actual inclusion follows from the fact that CAs can simulate polynomial-time many-one reductions by TMs in real-time.

Given a problem instance f, A evaluates f recursively. Without loss of generality, $f = \wedge(f_1 \in_{\mathrm{SAT}}, \vee(f_2 \in_{\mathrm{TAUT}}, f'))$, where f' is some other problem instance; other instances of $\mathrm{SAT}^{\wedge}\text{-}\mathrm{TAUT}^{\vee}$ are obtained by replacing f_1, f_2, or f' with a trivial formula (e.g., a trivial tautology).

To evaluate $f_1 \in_{\mathrm{SAT}}$, A emulates the behavior of the XCA for SAT (see Theorem 3); however, special care must be taken to ensure A does not halt prematurely. All computation branches retain a copy of f. Whenever a branch obtains a "true" result, the respective cells do not directly accept (as in the original construction); instead, they proceed with evaluating the formula's next connective. Conversely, if the result is false, the respective cells simply enter the reject state. The behavior for $f_2 \in_{\mathrm{TAUT}}$ is analogous, with A emulating the XCA for TAUT instead (and with exchanged accept and reject states, accordingly). Additionally, the accept and reject states are such that cells transition between them back and forth[1], and we (arbitrarily) enforce accept states only exist in even- and reject states in odd-numbered steps[2].

---

[1] That is, $\delta(\ell) = a$ for $\ell(0) = r$ and vice-versa, where δ and ℓ are as in Definition 2.

[2] This may be accomplished, for example, by using a bit counter in the cells' states, and having cells wait for a step before transitioning to an accept or reject state if needed.

If $f_1 \notin$ SAT, all branches of A transition into the reject state, and A rejects. Otherwise, f_1 is satisfiable; thus, at least one branch obtains a "true" result, and A continues to evaluate f until the (aforementioned) base case is reached. An analogous argument applies for f_2. Note the synchronicity of the branches guarantee they operate exactly the same and terminate at the same time. The repeated transition between accept and reject states guarantee the only cells relevant for the final decision of A are those in the branches which are still "active" (in the sense they are still evaluating f).

It is concluded that A accepts f if and only if it evaluates to true and rejects it otherwise. A runs in polynomial time (in $|f|$) since f has at most $|f|$ predicates and since evaluating a predicate requires polynomial time in $|f|$.   □

For the converse, we express an XCA computation as a $\text{SAT}^{\wedge}\text{-TAUT}^{\vee}$ instance. The main effort here lies on defining the appropriate "variables":

Definition 4 ($\text{STATE}_\forall$). *Let A be an XCA, and let V_A be the set of triples (w, t, z), w being an input for A, $t \in \{0,1\}^+$ a (standard) binary representation of $\tau \in \mathbb{N}_0$, and z a state of A. $\text{STATE}_\forall(A) \subseteq V_A$ is the subset of triples such that, if A is given w as input, then after τ steps all active cells are in state z.*

Lemma 2. *If A has polynomial time complexity, $\text{STATE}_\forall(A) \in \textbf{coNP}$.*

Proof. Fix an efficiently computable polynomial $p: \mathbb{N} \to \mathbb{N}_0$ such that A always terminates after at most $p(n)$ steps for an input of size n. Consider an NTM T which non-deterministically picks an active cell in step τ of A for input w, computes its state z' in polynomial time, and accepts if and only if $z' = z$. Additionally, suppose that, by doing so, T covers all active cells of A in step τ. Furthermore, to ensure T only simulates A for a polynomial number of steps (in $|w|$), T determines $p(|w|)$ and rejects in case $\tau > p(|w|)$; this is not a restriction because of the choice of p. The claim follows immediately from the existence of such a T: If all computation branches of T accept, then in step τ all cells of A are in state z; otherwise, there is a cell in a state which is not z, and T rejects. The rest of the proof is concerned with the construction of T as well as showing that its branches cover all active cells of A in its final configuration for the input w.

To compute the state of an active cell in step τ, T calculates a series of *subconfigurations* $c_0, \ldots, c_\tau$ of A, that is, contiguous excerpts of the global configuration of A. As the number of cells in an XCA may increase exponentially in the number of computation steps, bounding c_i is essential to ensure T runs in polynomial time; in particular, T maintains $|c_i| = 1 + 2(\tau - i)$, thus ensuring the lengths of the c_i are linear in τ (which, in turn, is polynomial in $|w|$). This choice of length for the c_i ensures each of the subconfigurations correspond to a cell of A surrounded by $\tau - i$ cells on either side.[3] The non-determinism of T is used precisely in picking the cells from c_i which are to be included in the next subconfiguration c_{i+1}.

---

[3] That is to say, each c_i corresponds to the so-called extended $(\tau - i)$-neighborhood of a cell of A.

The initial subconfiguration c_0 is set to be $q^{2\tau} w q^{2\tau}$, thus containing the input word as well as (as shall be proven) a sufficiently large number of surrounding quiescent cells. To obtain c_{i+1} from c_i, T applies the global transition function of A on c_i to obtain a new temporary subconfiguration c'_{i+1}. The next state of the two "boundary" cells (i.e, those belonging to indices 0 and $|c_i| - 1$) cannot be determined since the state of their neighbors is unknown; thus, they are excluded from c'_{i+1}. The same applies to any hidden cell which remains so. c'_{i+1} contains, as a result, $|c_i| - 2$ active cells from the previous configuration c_i in addition to a maximum of $|c_i| - 1$ previously hidden cells. To maintain $|c_i| = 1 + 2(\tau - i)$, T non-deterministically chooses a contiguous subset s of c'_{i+1} containing $1 + 2(\tau - i)$ cells and sets c_{i+1} to s; when doing so, T ignores subsets containing solely quiescent cells. That there are enough cells to choose from is, again, ensured by the fact that c'_{i+1} contains at least $|c_i| - 2$ active cells from the previous configuration c_i.

The process of selecting a next subconfiguration c_{i+1} from c_i is depicted in Fig. 2. In the figure, $|c_i|$ has been replaced with n for legibility. T at first applies the global transition function of A to obtain an intermediate subconfiguration c'_{i+1} with $m = |c'_{i+1}|$ cells. Because of hidden cells, c'_{i+1} may consist of $n - 2 \le m \le 2n - 3$ cells. Non-determinism is used to select a contiguous subconfiguration of $n - 2$ cells, thus giving rise to c_{i+1}.

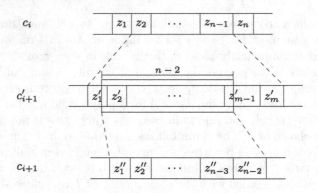

Fig. 2. Illustration of how T obtains the next subconfiguration c_{i+1} from c_i.

This concludes the construction of T. Note it runs in polynomial time since the invariant $|c_i| = 1 + 2(\tau - i)$ guarantees the number of states T computes in each step is bounded by a multiple of τ, which, as previously discussed, is bounded by $p(|w|)$. Only $|w|$ has to be taken into account when estimating the time complexity of T since the encoding of z is constant with respect to $|w|$ while that of t is logarithmic with respect to $p(|w|)$; expressing the problem instance (w, t, z) requires, as a result, asymptotically $|w|$ space.

To show all active cells of A in step τ are covered by T, it suffices to prove the following by induction: Let $i \in \{0, \ldots, \tau\}$, and let $z_1, \ldots, z_m$ be the active cells of A in step i; then T covers all subconfigurations of $q^{2(\tau-i)} z_1 \cdots z_m q^{2(\tau-i)}$ of size $1 + 2(\tau - i)$. Note this corresponds to T covering all subconfigurations of

A in step i which contain at least one active cell; thus, when T reaches step τ, it covers all subconfigurations of $z_1 \cdots z_m$ of size 1, that is, all active cells.

The induction basis follows from $c_0 = q^{2\tau} w q^{2\tau}$. For the induction step, fix a step $0 < i \leq \tau$ and assume the claim holds for all steps prior to i. To each subconfiguration of $q^{2(\tau-i)} z_1 \cdots z_m q^{2(\tau-i)}$ having size $1 + 2(\tau - i)$ corresponds a cell w which is located in its center; such subconfiguration is denoted by $s_i(w)$. Now let $s_i(w)$ be given, in which case there are three cases to be considered: w was active in step $i-1$; w was a hidden cell which became active in the transition to step i; or w was a quiescent cell in step $i - 1$ and, by $|s_i(w)| = 1 + 2(\tau - i)$, is at most $\tau - i$ cells away from z_1 or z_m.

In the first case, by the induction hypothesis, there is a value of c_{i-1} corresponding to $s_{i-1}(w)$; since only the two boundary cells are present in c_{i-1} but not in c_i', choosing c_i from c_i' with w as its middle cell yields $s_i(w)$. In the second, for any of the two parents p_1 and p_2 of w, there are, by the induction hypothesis, values of c_{i-1} which equal $s_{i-1}(p_1)$ and $s_{i-1}(p_2)$; in either case, choosing c_i from c_i' with w as its middle cell again yields $s_i(w)$.

Finally, if w was a quiescent cell, then, without loss of generality, consider the case in which w was located to the left of the active cells in step $i - 1$. By the induction hypothesis, for each cell w' up to $\tau - i + 1$ cells away from the leftmost active cell z_1 there is a value of c_{i-1} corresponding to $s_{i-1}(w')$, and the first case applies; the only exception is if c_i would then contain only quiescent cells, in which case w would be located strictly more than $\tau - i$ cells away from z_1, thus contradicting our previous assumption. The claim follows.   □

The following proposition completes our argument by reduction:

Proposition 2. XCAP $\subseteq \leq_{tt}^p(\mathbf{NP})$.

Proof. Let $L \in \mathbf{XCAP}$, and let A be an XCA for L whose time complexity is bounded by a polynomial $p \colon \mathbb{N} \to \mathbb{N}_0$. Additionally, let w be an input for A, V_A be as in Definition 4, and let $V = V_A \cdot \{\models_{\mathrm{STATE}_{\forall}(A)}\}$, where $\models_{\mathrm{STATE}_{\forall}(A)}$ is a syntactic symbol standing for membership in $\mathrm{STATE}_{\forall}(A)$ (cf. Definition 1). Define $f_0(w), \ldots, f_{p(n)}(w) \in \mathrm{BOOL}_V$ recursively as follows:

$$f_i(w) := \begin{cases} \vee(\,(w, i, a)\,\models_{\mathrm{STATE}_{\forall}(A)}, \wedge(\,\neg(\,(w, i, r)\,\models_{\mathrm{STATE}_{\forall}(A)}\,), f_{i+1}\,)\,), & i \neq p(n) \\ \vee(\,(w, p(n), a)\,\models_{\mathrm{STATE}_{\forall}(A)}, \neg(\,(w, p(n), r)\,\models_{\mathrm{STATE}_{\forall}(A)}\,)\,), & i = p(n) \end{cases}$$

Lemma 2 together with the **coNP**-completeness of TAUT (see Theorem 1) ensures each subformula of the form $(w, i, a)\,\models_{\mathrm{STATE}_{\forall}(A)}$ is polynomial-time many-one reducible to an equivalent[4] $\mathrm{SAT}^{\wedge}\text{-}\mathrm{TAUT}^{\vee}$ formula $g \models_{\mathrm{TAUT}}$, g being a TAUT instance. Similarly, each subformula $\neg((w, i, a)\,\models_{\mathrm{STATE}_{\forall}(A)})$ is reducible to an equivalent formula $h \models_{\mathrm{SAT}}$. Since each of the $f_i(w)$ may contain only polynomially (respective to $|w|$) many connectives, each is polynomial-time (many-one) reducible to an equivalent $\mathrm{SAT}^{\wedge}\text{-}\mathrm{TAUT}^{\vee}$ instance $f_i'(w)$.

---

[4] In the sense of evaluating to the same truth value under the respective interpretations (see Definition 1).

By the definition of XCA (i.e., Definitions 2 and 3) and our choice of p, $f'(w) := f'_0(w)$ is true if and only if A accepts w. Since $f'(w)$ is such that $|f'(w)|$ is polynomial in $|w|$, this provides a polynomial-time (many-one) reduction of L to a problem instance of $\text{SAT}^\wedge\text{-TAUT}^\vee \in \leq_{tt}^p(\mathbf{NP})$. The claim follows. $\square$

4   Immediate Implications

This section covers some immediate corollaries of Theorem 4 regarding XCA variants. In particular, we address XCAs with multiple accept and reject states, followed by XCAs with acceptance conditions differing from that from Definition 3, in particular the two other classical acceptance conditions for CAs [16].

4.1   XCAs with Multiple Accept and Reject States

Recall the definition of an XCA specifies a single accept and a single reject state (see Sect. 2.2). Consider XCAs with multiple accept and reject states. As shall be proven, the respective polynomial-time class (**MAR-XCAP**) remains equal to **XCAP**. In the case of TMs, the equivalent result (i.e., TMs with a single accept and a single reject state are as efficient as standard TMs) is trivial, but such is not the case for XCAs. Recall the acceptance condition of an XCA requires orchestrating the states of multiple, possibly exponentially many cells. In addition, an XCA with multiple accept states may, for instance, attempt to accept whilst saving its current state (i.e., a cell in state z may assume an accept state a_z while simultaneously saving state z). Such is not the case for standard XCAs (i.e., as specified in Definition 3), in which all accepting cells have necessarily the same state.

Definition 5 (MAR-XCA, MAR-XCAP). *A multiple accept-reject XCA (MAR-XCA) D is an XCA with state set Q and which admits subsets $A, R \subseteq Q$ of accept and reject states, respectively. D accepts (rejects) if its active cells all have states in A (R), and it halts upon accepting or rejecting. In addition, D is required to either accept or reject its input after a finite number of steps.* **MAR-XCAP** *denotes the MAR-XCA analogue of* **XCAP**.

The following generalizes $\text{STATE}_\forall$ (see Definition 4 and Lemma 2) to the case of MAR-XCAs:

Definition 6 (STATE$_\forall^{\text{MAR}}$). *Let A be a MAR-XCA with state set Q, and let V_A be the set of triples (w, t, Z), w being an input for A, $t \in \{0,1\}^+$ a binary encoding of $\tau \in \mathbb{N}_0$, and $Z \subseteq Q$. $\text{STATE}_\forall^{\text{MAR}}(A) \subseteq V_A$ is the subset of triples such that, if A is given w as input, after t steps all active cells have states in Z.*

Lemma 3. *If A has polynomial time complexity,* $\text{STATE}_\forall^{\text{MAR}} \in \mathbf{coNP}$.

Proof. Simply adapt the NTM from the proof of Lemma 2 so as to accept if and only if the final state is contained in Z. $\square$

Proceeding as in the proof of Proposition 2 (simply using $\text{STATE}_\forall^{\text{MAR}}$ instead of $\text{STATE}_\forall$) yields:

Theorem 5. MAR-XCAP = XCAP.

Proof. Define formulas $f_i(w)$ as in the proof of Proposition 2 while replacing $\text{STATE}_\forall$ with $\text{STATE}_\forall^{\text{MAR}}$, the accept state a with the set A, and the reject state r with the set R. Lemma 3 guarantees the reductions to $\text{SAT}^\wedge\text{-TAUT}^\vee$ are all efficient. This implies **MAR-XCAP** $\subseteq \leq_{tt}^p(\textbf{NP}) = $ **XCAP**. Since MAR-XCAs are a generalization of XCAs, the converse inclusion is trivial. $\square$

4.2   Existential XCA

The remainder of this section is concerned with XCAs variants which use the two other acceptance conditions from [6,8,16,20]. The first is that of a single final state being present in the CA's configuration sufficing for termination.

Definition 7 (EXCA, EXCAP). *An* existential[5] *XCA (EXCA) is an XCA with the following acceptance condition: If at least one of its cells is in the accept (reject) state a (r), then the EXCA accepts (rejects). The coexistence of accept and reject states in the same global configuration is disallowed (and any machine contradicting this requirement is, by definition, not an EXCA).* **EXCAP** *denotes the EXCA analogue of* **XCAP**.

Disallowing the coexistence of accept and reject states in the global configuration of an EXCA is necessary to ensure a consistent accepting behavior. An alternative would be to establish a priority relation between the two (e.g., an accept state overrules a reject one); nevertheless, this behavior can be emulated by our chosen variant with only constant delay. This is accomplished by introducing binary counters to delay state transitions and assure, for instance, that accept and reject states exist only in even- and odd-numbered steps, respectively.

Despite the diverse accepting behavior, the following holds for EXCAs:

Theorem 6. EXCAP = XCAP = $\leq_{tt}^p(\textbf{NP})$.

Note this is an equivalence between two disparately complex acceptance behaviors: As by Definition 3, in an XCA all cells must agree on the final decision; on the other hand, in an EXCA, a single, arbitrary cell suffices. We ascribe this phenomenon to **XCAP** $= \leq_{tt}^p(\textbf{NP})$ being equal to its complementary class.

As for the proof of Theorem 6, first note that Proposition 1 may easily be restated in the context of EXCAs:

Proposition 3. $\leq_{tt}^p(\textbf{NP}) \subseteq $ EXCAP.

Proof. Consider the XCA A for $\text{SAT}^\wedge\text{-TAUT}^\vee$ from the proof of Proposition 1; by adapting it, we obtain an EXCA B which decides $\text{TAUT}^\wedge\text{-SAT}^\vee$ in polynomial time. $\text{TAUT}^\wedge\text{-SAT}^\vee$ is the problem analogous to $\text{SAT}^\wedge\text{-TAUT}^\vee$

---

[5] An allusion to the existential states of alternating Turing machines (ATMs).

and which is obtained by exchanging "TAUT" and "SAT" in Definition 1. As $\text{SAT}^\wedge\text{-TAUT}^\vee$, $\text{TAUT}^\wedge\text{-SAT}^\vee$ is $\leq_{tt}^p(\mathbf{NP})$-complete (see Theorem 2).

To evaluate a predicate of the form $f \Subset_{\text{TAUT}}$, B proceeds as A and emulates the behavior of the XCA deciding TAUT (see Theorem 3); however, unlike A, the computation branches of B which evaluate to false reject immediately while it is those that evaluate to true that continue evaluating the input formula. As a result, if $f \in \text{TAUT}$, all branches of B evaluate to true and continue evaluating the input in a synchronous manner; otherwise, there is a branch evaluating to false, and, since a single rejecting cell suffices for it to reject, B rejects immediately. The evaluation of $f \Subset_{\text{SAT}}$ is carried out analogously.

The modifications to A to obtain B do not impact its time complexity whatsoever; thus, B also has polynomial time complexity. $\qquad\qquad\Box$

For the converse inclusion, consider the following **NP** analogue of the $\text{STATE}_\forall$ language (cf. Definition 4 and Lemma 2):

Definition 8 (STATE∃). *Let A be an XCA, and let V be the set of triples (w, t, z) as in Definition 4. $\text{STATE}_\exists \subseteq V$ is the subset of triples such that, for the input w, after t steps at least one of the active cells of A is in state z.*

Lemma 4. *If A has polynomial time complexity, $\text{STATE}_\exists(A) \in \mathbf{NP}$.*

Proof. Consider the NTM T from Lemma 2 and notice that, if any of the active cells of A in step τ have state z, then T will have at least one accepting branch; otherwise, none of the active cells of A in step τ have state z; thus, all branches of T are rejecting. $\qquad\qquad\Box$

Using Lemma 4 to proceed as in Proposition 2 yields the following, from which Theorem 6 follows:

Proposition 4. EXCAP $\subseteq \leq_{tt}^p(\mathbf{NP})$.

4.3   One-Cell-Decision XCA

We turn to the discussion of XCAs whose acceptance condition is defined in terms of a distinguished cell which directs the automaton's decision, considered the standard acceptance condition for CAs [9]. This behavior is similar to the existential variant in the sense that a single cell suffices to trigger the automaton's termination; the difference lies in the position of this single cell being immutable.

We consider only the case in which the decision cell is the leftmost active cell in the initial configuration (i.e., cell 0). By a *one-cell-decision XCA (1XCA)* we refer to an XCA which accepts if and only if 0 is in the accept state and rejects if and only if cell zero is in the reject state. Let **1XCAP** denote the polynomial-time class of 1XCAs.

The position of the decision cell is fixed; with a polynomial-time restriction in place, it can only communicate with cells which are a polynomial (in the length of the input) number of steps apart. As a result, despite a 1XCA being able to efficiently increase its number of active cells exponentially (see Lemma 1), any cells impacting its decision behavior must be at most a polynomial number of cells away from the decision cell. Thus:

Theorem 7. 1XCAP = P.

Proof. The inclusion **1XCAP** $\supseteq$ **P** is trivial. For the converse, recall the construction of the NTM T in Lemma 2. T can be modified such that it works deterministically and always chooses the next configuration c_{i+1} from c_i by selecting cell zero as the middle cell. If cell zero is accepting, then T accepts immediately; if it is rejecting, then T also rejects immediately. This yields a simulation of a 1XCA by a (deterministic) TM which is only polynomially slower, thus implying **1XCAP** $\subseteq$ **P**.   □

5   Conclusion

This paper summarized the results of [13], which, in turn, expanded on the complexity-theoretic aspects of XCAs from [14]. The main result was the characterization **XCAP** $= \leq_{tt}^{p}(\mathbf{NP})$ in Sect. 3.1. In Sect. 4, XCAs with multiple accept and reject states were shown to be equivalent to the original model. Also in Sect. 4, two other variants based on varying acceptance conditions were considered: the existential (EXCA), in which a single, though arbitrary cell may direct the automaton's response; and the one-cell-decision XCA (1XCA), in which a fixed cell does so. In the first case, it was shown that the polynomial-time class **EXCAP** equals **XCAP**; in the latter, it was shown that the polynomial-time class **1XCAP** of 1XCAs equals **P**.

This paper has covered some XCA variants with diverse acceptance conditions. A topic for future work might be considering further variations in this sense (e.g., XCAs whose acceptance condition is based on *majority* instead of *unanimity*). Another avenue of research lies in restricting the capabilities of XCAs and analyzing the effects thereof (e.g., restricting 1XCAs or SXCAs to a polynomial number of cells). A final open question is determining what polynomial speedups, if any, 1XCAs provide with respect to 1CAs.

Acknowledgments. The author would like to thank Thomas Worsch for his mentoring, encouragement, and support during the writing of this paper. The author would also like to thank Dennis Hofheinz for pointing out a crucial mistake in a preliminary version of this paper as well as the anonymous referees for their valuable remarks and suggestions.

References

1. Arrighi, P., Dowek, G.: Causal graph dynamics. Inf. Comput. **223**, 78–93 (2013)
2. Buss, S.R., Hay, L.: On truth-table reducibility to SAT. Inf. Comput. **91**(1), 86–102 (1991)

3. Cook, S.A.: The complexity of theorem-proving procedures. In: Proceedings of the Third Annual ACM Symposium on Theory of Computing, STOC 1971, Shaker Heights, Ohio, USA, pp. 151–158. ACM (1971)
4. Dantchev, S.S.: Dynamic neighbourhood cellular automata. In: Visions of Computer Science - BCS International Academic Conference, Imperial College, London, UK, 22–24 September 2008, pp. 60–68 (2008)
5. Dubacq, J.-C.: Different kinds of neighborhood-varying cellular automata. Maîtrise/honour bachelor degree, École normale superiéure de Lyon (1994)
6. Ibarra, O.H., et al.: Fast parallel language recognition by cellular automata. Theor. Comput. Sci. **41**, 231–246 (1985)
7. Ilachinski, A., Halpern, P.: Structurally dynamic cellular automata. Complex Syst. **1**(3), 503–527 (1987)
8. Kim, S., McCloskey, R.: A characterization of constant-time cellular automata computation. Physica D **45**(1–3), 404–419 (1990)
9. Kutrib, M.: Cellular automata and language theory. In: Encyclopedia of Complexity and Systems Science, pp. 800–823 (2009)
10. Kutrib, M., Malcher, A., Wendlandt, M.: Shrinking one-way cellular automata. In: Kari, J. (ed.) AUTOMATA 2015. LNCS, vol. 9099, pp. 141–154. Springer, Heidelberg (2015). https://doi.org/10.1007/978-3-662-47221-7_11
11. Ladner, R.E., Lynch, N.A.: Relativization of questions about log space computability. Math. Syst. Theory **10**, 19–32 (1976)
12. Ladner, R.E., et al.: A comparison of polynomial time reducibilities. Theor. Comput. Sci. **1**(2), 103–123 (1975)
13. Modanese, A.: Complexity-theoretical aspects of expanding cellular automata. Master's thesis, Karlsruhe Institute of Technology (2018)
14. Modanese, A.: Shrinking and expanding one-dimensional cellular automata. Bachelor's thesis, Karlsruhe Institute of Technology (2016)
15. Modanese, A., Worsch, T.: Shrinking and expanding cellular automata. In: Cook, M., Neary, T. (eds.) AUTOMATA 2016. LNCS, vol. 9664, pp. 159–169. Springer, Cham (2016). https://doi.org/10.1007/978-3-319-39300-1_13
16. Rosenfeld, A.: Picture Languages. Academic Press, New York (1979)
17. Rosenfeld, A., Wu, A.Y.: Reconfigurable cellular computers. Inf. Control **50**(1), 60–84 (1981)
18. Rosenfeld, A., et al.: Fast language acceptance by shrinking cellular automata. Inf. Sci. **30**(1), 47–53 (1983)
19. Smith III, A.R.: Simple computation-universal cellular spaces. J. ACM **18**(2), 339–353 (1971)
20. Sommerhalder, R., van Westrhenen, S.C.: Parallel language recognition in constant time by cellular automata. Acta Inf. **19**, 397–407 (1983)
21. Tomita, K., et al.: Graph automata: natural expression of self-reproduction. Phys. D: Nonlinear Phenom. **171**(4), 197–210 (2002)
22. Wagner, K.W.: Bounded query classes. SIAM J. Comput. **19**(5), 833–846 (1990)

Iterative Arrays with Finite Inter-cell Communication

Martin Kutrib and Andreas Malcher[✉]

Institut für Informatik, Universität Giessen, Arndtstr. 2, 35392 Giessen, Germany
{kutrib,andreas.malcher}@informatik.uni-giessen.de

Abstract. Iterative arrays whose internal inter-cell communication is quantitatively restricted are investigated. The quantity of communication is measured by counting the number of uses of the links between cells. In particular, iterative arrays are studied where the maximum number of communications per cell occurring in accepting computations is drastically bounded by a constant number. Additionally, the iterative arrays have to work in realtime. We study the computational capacity of such devices. One main result is that a strict and dense hierarchy with respect to the constant number of communications exists. Due to their very restricted communication, the question arises whether the usually studied decidability problems such as, for example, emptiness, finiteness, inclusion, or equivalence become decidable for such devices. However, by reduction of Hilbert's tenth problem it can be shown that all such decidability questions remain undecidable.

1 Introduction

Devices of homogeneous, interconnected, parallel acting automata have widely been investigated from a computational capacity point of view. Multidimensional devices with nearest neighbor connections whose cells are finite automata are commonly called *cellular automata* (CA). The cells work synchronously at discrete time steps. If the input mode is sequential to a distinguished communication cell, such devices are called *iterative arrays*.

In connection with formal language recognition one-dimensional iterative arrays (IA) have been introduced in [3], where it was shown that the language family accepted by realtime IAs forms a Boolean algebra not closed under concatenation and reversal. In [2] it is shown that for every context-free grammar a two-dimensional lineartime iterative array parser exists. A realtime IA for prime numbers has been constructed in [4]. A characterization of various types of IAs in terms of restricted Turing machines and several results, especially speed-up theorems, are given in [6,7]. Several more results concerning formal languages can be found, for example, in the survey [9].

It is obvious that inter-cell communication is an essential resource for iterative arrays and can be measured qualitatively as well as quantitatively. In the first case, the number of different messages to be communicated by the cells is

© IFIP International Federation for Information Processing 2019
Published by Springer Nature Switzerland AG 2019
A. Castillo-Ramirez and P. P. B. de Oliveira (Eds.): AUTOMATA 2019, LNCS 11525, pp. 35–47, 2019.
https://doi.org/10.1007/978-3-030-20981-0_3

bounded by some fixed constant. IAs with this restricted inter-cell communication have been investigated in [18,19] with respect to the algorithmic design of sequence generation. In particular, it is shown that several infinite, non-regular sequences such as exponential or polynomial, Fibonacci, and prime sequences can be generated in realtime. In connection with language recognition and decidability questions multi-dimensional IAs and one-dimensional (one-way) CAs with restricted communication have intensively been studied in [10,14,20].

For a quantitative measure of communication in iterative arrays the number of uses of the links between cells is counted. Additionally, it is distinguished between bounds on the *sum* of all communications of an accepting computation and bounds on the *maximum number* of communications *per cell* occurring in accepting computations. There are quite a few results in the literature with respect to these measures. Results for (one-way) cellular automata may be found in [12,13]. In [11,13] also cellular automata are investigated that are restricted with respect to the qualitative *and* the quantitative measure. The main results are in both cases hierarchy results and the undecidability of almost all commonly studied decidability questions such as emptiness, finiteness, equivalence, inclusion, regularity, and context-freeness. It should be noted that already a finite amount of communication per cell is sufficient to obtain undecidability results for cellular automata. First results on iterative arrays with restricted communication are presented in [15] and comprise again hierarchy results as well as undecidability results for the above questions. Concerning the measure on the maximum communication per cell the undecidability results hold as long as at least a logarithmic number of communications per cells is allowed. Moreover, it is stated as an open question whether the undecidable questions become decidable when the allowed communication is even more restricted, namely, to be bounded by a constant number.

In this paper, we can answer the latter question negatively. In addition, we establish a strict and dense hierarchy with respect to the constant number of communications. The paper is organized as follows. In the next section, we present some basic notions and definitions, introduce the classes of max communication bounded iterative arrays, and give an illustrative example. Then, in Sect. 3 we show that for every $k \geq 2$, IAs with at most $k+1$ communications per cell are more powerful than devices with at most k communications per cell. For $k \in \{0,1,2\}$ it turns out that devices with at most k communications per cell can accept regular languages only. Section 4 is devoted to showing the undecidability of the usually studied decidability questions for IAs working in realtime with a constant number of communications per cell. This is done by a reduction of Hilbert's tenth problem and requires a couple of consecutive constructions of IAs with a constant number of communications per cell, whereby the goal is to evaluate a polynomial with integral coefficients given by an instance of Hilbert's tenth problem.

2 Definitions and Preliminaries

We denote the non-negative integers by $\mathbb{N}$. Let Σ denote a finite set of letters. Then we write Σ^* for the *set of all finite words* (strings) consisting of letters from Σ. The *empty word* is denoted by λ, and we set $\Sigma^+ = \Sigma^* \setminus \{\lambda\}$. A subset of Σ^* is called a *language* over Σ. For the *reversal of a word* w we write w^R and for its *length* we write $|w|$. For the number of occurrences of a symbol x in w we use the notation $|w|_x$. A language L over some alphabet $\{a_1, a_2, \ldots, a_k\}$ is said to be *letter bounded*, if $L \subseteq a_1^* a_2^* \cdots a_k^*$. In general, we use $\subseteq$ for *inclusions* and $\subset$ for *strict inclusions*.

A one-dimensional iterative array is a linear, semi-infinite array of identical deterministic finite state machines, sometimes called cells. Except for the leftmost cell each one is connected to its both nearest neighbors (see Fig. 1). For convenience we identify the cells by their coordinates, that is, by non-negative integers. The distinguished leftmost cell at the origin is connected to its right neighbor and, additionally, equipped with a one-way read-only input tape. At the outset of a computation the input is written on the input tape with an infinite number of end-of-input symbols to the right, and all cells are in the so-called quiescent state. The finite state machines work synchronously at discrete time steps. The state transition of all cells but the communication cell depends on the current state of the cell itself and on the information which is currently sent by its neighbors. The information sent by a cell depends on its current state and is determined by so-called communication functions. The state transition of the communication cell additionally depends on the input symbol to be read next. The head of the one-way input tape is moved to the right in each step. With an eye towards recognition problems the machines have no extra output tape but the states are partitioned into accepting and rejecting states.

Formally, an *iterative array* (IA) is a system $\langle S, F, A, B, \triangledown, s_0, b_l, b_r, \delta, \delta_0 \rangle$, where S is the finite, nonempty set of *cell states*, $F \subseteq S$ is the set of *accepting states*, A is the finite set of *input symbols*, B is the finite set of *communication symbols*, $\triangledown \notin A$ is the *end-of-input symbol*, $s_0 \in S$ is the *quiescent state*, $b_l, b_r : S \to B \cup \{\bot\}$ are *communication functions* which determine the information to be sent to the left and right neighbors, where $\bot$ means nothing to send and $b_l(s_0) = b_r(s_0) = \bot$, $\delta : (B \cup \{\bot\}) \times S \times (B \cup \{\bot\}) \to S$ is the *local transition function for all but the communication cell* satisfying $\delta(\bot, s_0, \bot) = s_0$, and $\delta_0 : (A \cup \{\triangledown\}) \times S \times (B \cup \{\bot\}) \to S$ is the *local transition function for the communication cell*.

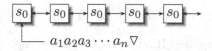

Fig. 1. Initial configuration of an iterative array.

Let M be an IA. A configuration of M at some time $t \geq 0$ is a description of its global state which is a pair (w_t, c_t), where $w_t \in A^*$ is the remaining input sequence and $c_t : \mathbb{N} \to S$ is a mapping that maps the single cells to their current states. The configuration (w_0, c_0) at time 0 is defined by the input word w_0 and the mapping c_0 that assigns the quiescent state to all cells, while subsequent configurations are chosen according to the global transition function Δ that is induced by δ and δ_0 as follows: Let (w_t, c_t), $t \geq 0$, be a configuration. Then its successor configuration $(w_{t+1}, c_{t+1}) = \Delta(w_t, c_t)$ is as follows.

$$c_{t+1}(i) = \delta(b_r(c_t(i-1)), c_t(i), b_l(c_t(i+1)))$$

for all $i \geq 1$, and $c_{t+1}(0) = \delta_0(a, c_t(0), b_l(c_t(1)))$, where $a = \triangledown$ and $w_{t+1} = \lambda$ if $w_t = \lambda$, as well as $a = a_1$ and $w_{t+1} = a_2 a_3 \cdots a_n$ if $w_t = a_1 a_2 \cdots a_n$.

We remark that we obtain the classical definition of IA, if we set $B = S$ and $b_l(s) = b_r(s) = s$ for all $s \in S$.

An input w is accepted by an IA M if at some time i during the course of its computation the communication cell enters an accepting state. The *language accepted by* M is denoted by $L(M)$. Let $t : \mathbb{N} \to \mathbb{N}$, $t(n) \geq n + 1$ be a mapping. If all $w \in L(M)$ are accepted with at most $t(|w|)$ time steps, then M and $L(M)$ are said to be of time complexity t.

The family of all languages which are accepted by some type of device X with time complexity t is denoted by $\mathscr{L}_t(X)$. If t is the function $n + 1$, acceptance is said to be in *realtime* and we write $\mathscr{L}_{rt}(X)$. Since for nontrivial computations an IA has to read at least one end-of-input symbol, realtime has to be defined as $(n+1)$-time. The *lineartime* languages $\mathscr{L}_{lt}(X)$ are defined according to $\mathscr{L}_{lt}(X) = \bigcup_{r \in \mathbb{Q}, r \geq 1} \mathscr{L}_{r \cdot n}(X)$.

In the following we study the impact of communication in iterative arrays. The communication is measured by the number of uses of the links between cells. It is understood that whenever a communication symbol not equal to $\perp$ is sent, a communication takes place. Here we do not distinguish whether either or both neighboring cells use the link. More precisely, the number of communications between cell i and cell $i + 1$ up to time step t is defined by

$$\mathrm{com}(i, t) = |\{\, j \mid 0 \leq j < t \text{ and } (b_r(c_j(i)) \neq \perp \text{ or } b_l(c_j(i+1)) \neq \perp)\,\}| \,.$$

For computations we now consider the maximal number of communications between two cells. Let $c_0, c_1, \ldots, c_{t(|w|)}$ be the sequence of configurations computed on input w by some iterative array with time complexity $t(n)$, that is, the *computation on* w. Then we define

$$\mathrm{mcom}(w) = \max\{\, \mathrm{com}(i, t(|w|)) \mid 0 \leq i \leq t(|w|) - 1 \,\}.$$

Let $f : \mathbb{N} \to \mathbb{N}$ be a mapping. If all $w \in L(M)$ are accepted with computations where $\mathrm{mcom}(w) \leq f(|w|)$, then M is said to be *max communication bounded by* f. We denote the class of IA that are max communication bounded by some function f by $\mathrm{MC}(f)$-IA. In addition, we use the notation *const* for functions from $O(1)$.

To illustrate the definitions we start with an example. In the next section it turns out that the non-regular language $\{\,a^n b^n \mid n \geq 1\,\}$ is accepted by some realtime iterative array using at most three communications per inter-cell link. The following example reveals that only five communications per inter-cell link are sufficient to accept the non-semilinear and, hence, non-context-free language $L = \{\,a^{3n+2\lfloor\sqrt{n}\rfloor} b^{2n} \mid n \geq 1\,\}$ in realtime. Moreover, L is a subset of $a^* b^*$ and, hence, a letter-bounded language.

Example 1. The language L belongs to $\mathscr{L}_{rt}(\mathrm{MC}(5)\text{-}\mathrm{IA})$.

The basic idea is to use the construction given in [17] where a cellular automaton is described such that the nth cell enters a designated state s exactly at time step $2n + \lfloor\sqrt{n}\rfloor$. We basically implement this construction, but realize it with speed $1/2$ in contrast to the construction in [17]. Hence, the nth cell enters a designated state exactly at time step $4n + 2\lfloor\sqrt{n}\rfloor$. When the communication cell reads the first b, it sends a signal with maximum speed to the right which arrives in the nth cell exactly at the moment when state s should be entered. In this case, another signal is sent with maximum speed to the left and the input is accepted if the latter signal reaches the communication cell when the end-of-input symbol is read. Thus, L is accepted by a realtime IA. Moreover, the construction given in [17] needs three right-moving signals. Thus, four right-moving signals and one left-moving signal are sufficient to accept L. This shows that the IA constructed is a realtime MC(5)-IA.   ∎

3 $k + 1$ Communications Are Better than k

This section is devoted to studying the impact of the precise finite number k of communications between cells. It turns out that this number in fact matters unless it is very small. That is, we will obtain an infinite strict hierarchy for $k \geq 2$, whereas the families of languages accepted with 0, 1, and 2 communications per inter-cell link coincide with the regular languages. We start at the bottom of the hierarchy.

Proposition 2. *The language families $\mathscr{L}_{rt}(MC(0)\text{-}IA)$, $\mathscr{L}_{rt}(MC(1)\text{-}IA)$, and $\mathscr{L}_{rt}(MC(2)\text{-}IA)$ coincide with the family of regular languages.*

Proof. The proof is trivial for MC(0)-IAs. In this case the iterative array has the computational capacity of the communication cell, that is, of a deterministic finite automaton. Similarly, the proof is obvious for MC(1)-IAs. The sole communication on the inter-cell link between the communication cell and its neighbor sends some information to the right, but this information can never come back to the communication cell which, thus, is not affected by the communication at all. We conclude that the computational capacity also for MC(1)-IAs is that of deterministic finite automata.

Let us now consider MC(2)-IAs. Let the first communication on the inter-cell link between the communication cell and its neighbor take place at time step $t \geq 0$. Before, all cells to the right of the communication cell are quiescent. So,

the information transmitted by the communication causes the quiescent array
to perform some computation and possibly to transmit some information back
to the communication cell. More communications of the communication cell are
useless. The computation performed by the array to the right of the communication cell only depends on the information transmitted during the first communication. However, these finitely many cases can be precomputed. Implementing
the transition function of the communication cell so that it simulates the computations of the array in some state register allows to safely remove the first
communication. In this way we obtain an equivalent MC(1)-IA that accepts regular languages only. Needless to say that a second communication from left to
right is useless and can be omitted as well.   □

The witnesses for the hierarchy are languages whose words are repetitions of
unary blocks of the same size but with alternating symbols. For $i \geq 2$ define

$$L_{\mathrm{hi}} = \begin{cases} \{(a^n b^n)^{\frac{i}{2}} \mid n \geq 1\} & \text{if } i \text{ is even} \\ \{(a^n b^n)^{\lfloor \frac{i}{2} \rfloor} a^n \mid n \geq 1\} & \text{if } i \text{ is odd} \end{cases}.$$

Lemma 3. *For $i \geq 2$, the language L_{hi} is accepted by some $MC(i+1)$-IA.*

Theorem 4. *For $i \geq 2$, the family $\mathscr{L}_{rt}(MC(i)$-$IA)$ is strictly included in the
family $\mathscr{L}_{rt}(MC(i+1)$-$IA)$*

Proof. For $i \geq 2$, the witness language L_{hi} is used to show the strict inclusion
$\mathscr{L}_{rt}(MC(i)$-$IA) \subset \mathscr{L}_{rt}(MC(i+1)$-$IA)$. By Lemma 3, language L_{hi} is accepted
by some $MC(i+1)$-IA in realtime. So, it remains to be shown that L_{hi} is not
accepted by any $MC(i)$-IA in realtime.

Let $M = \langle S, F, A, B, \triangledown, s_0, b_l, b_r, \delta, \delta_0 \rangle$ be an iterative array accepting L_{hi}
in realtime. We consider the communications on the inter-cell link between the
communication cell and its neighbor. For clearer writing, we represent a configuration by a pair $(w, s\hat{c})$, where $w \in A^*$ is the remaining input sequence as usual,
$s \in S$ is the state of the communication cell, and $\hat{c}$ is a mapping that maps cells
$j \geq 1$ to their current states.

A key observation is that on unary inputs long enough the communication
cell will run into state cycles unless a communication takes place. So, let $w \in L_{\mathrm{hi}}$
be an accepted word whose block length n is long enough.

Assume that no communication takes place while processing the first $|S|$ sym-
bols a from the first block. Recall that by definition we have $b_l(s_0) = b_r(s_0) = \perp$.
Let $w = a^n v$ with the first letter in v being b. Then on processing the input pre-
fix $a^{|S|}$ the communication cell necessarily enters some state at least twice. That
is, there are $0 \leq p_1$, $1 \leq p_2$ with $p_1 + p_2 \leq |S|$ such that $\Delta^{p_1}(a^n v, s_0 \hat{c}_0) =
(a^{n-p_1} v, s_1 \hat{c}_0)$ and $\Delta^{p_2}(a^{n-p_1} v, s_1 \hat{c}_0) = (a^{n-p_1-p_2} v, s_1 \hat{c}_0)$. That is, there is a
state cycle of length p_2 leading from state s_1 to state s_1. So, the input $a^{n+p_2} v$
is accepted as well. From the contradiction it follows that there is at least one
communication during the first $|S|$ time steps.

Next, we consider the sub-computations that process the last $|S|$ symbols of
a block and the first $|S|^2 + |S|$ symbols of the following block. Without loss of

generality, let this factor be $a^{|S|}b^{|S|^2+|S|}$, and so $w = ua^{|S|}b^{|S|^2+|S|}v$. Assume that no communication takes place (on the inter-cell link between the communication cell and its neighbor) while processing this factor. Let

$$\Delta^{|u|}(ua^{|S|}b^{|S|^2+|S|}v, s_0\hat{c}_0) = (a^{|S|}b^{|S|^2+|S|}v, s\hat{c}).$$

Then there are $0 \le p_1$, $1 \le p_2$ with $p_1 + p_2 \le |S|$ such that

$$\Delta^{p_1}(a^{|S|}b^{|S|^2+|S|}v, s\hat{c}) = (a^{|S|-p_1}b^{|S|^2+|S|}v, s_1\hat{c}_1) \text{ and}$$
$$\Delta^{p_2}(a^{|S|-p_1}b^{|S|^2+|S|}v, s_1\hat{c}_1) = (a^{|S|-p_1-p_2}b^{|S|^2+|S|}v, s_1\hat{c}_2).$$

That is, there is a state cycle of length p_2 leading from state s_1 to state s_1.Continuing the computation without communication yields the existence of $0 \le p_3$, $1 \le p_4 \le |S|$ with $|S| \le p_1 + p_2 + p_3$ and $p_1 + p_2 + p_3 + p_4 \le 2|S|$ such that

$$\Delta^{p_3}(a^{|S|-p_1-p_2}b^{|S|^2+|S|}v, s_1\hat{c}_2) = (b^{|S|^2+2|S|-p_1-p_2-p_3}v, s_2\hat{c}_3),$$
$$\Delta^{p_4}(b^{|S|^2+2|S|-p_1-p_2-p_3}v, s_2\hat{c}_3) = (b^{|S|^2+2|S|-p_1-p_2-p_3-p_4}v, s_2\hat{c}_4), \text{ and}$$
$$\Delta^{p_2p_4}(b^{|S|^2+2|S|-p_1-p_2-p_3-p_4}v, s_2\hat{c}_4) = (b^{|S|^2+2|S|-p_1-p_2-p_3-p_4-p_2p_4}v, s_2\hat{c}_5).$$

That is, there is a state cycle of length p_4 leading from state s_2 to state s_2. Since there are no communications on the factor $a^{|S|}b^{|S|^2+|S|}$, we can replace this factor by the factor $a^{|S|+p_2p_4}b^{|S|^2+|S|-p_2p_4}$ of the same length, and the configurations to the right of the communication cell develop exactly as before. We obtain the sub-computations

$$\Delta^{p_1}(a^{|S|+p_2p_4}b^{|S|^2+|S|-p_2p_4}v, s\hat{c}) = (a^{|S|+p_2p_4-p_1}b^{|S|^2+|S|-p_2p_4}v, s_1\hat{c}_1) \text{ and}$$
$$\Delta^{p_2}(a^{|S|+p_2p_4-p_1}b^{|S|^2+|S|-p_2p_4}v, s_1\hat{c}_1) = (a^{|S|+p_2p_4-p_1-p_2}b^{|S|^2+|S|-p_2p_4}v, s_1\hat{c}_2).$$

Since the communication cell runs through the state cycle of length p_2 leading from state s_1 to state s_1 and there is no communication, the sub-computation continues as

$$\Delta^{p_2p_4}(a^{|S|+p_2p_4-p_1-p_2}b^{|S|^2+|S|-p_2p_4}v, s_1\hat{c}_2) = (a^{|S|-p_1-p_2}b^{|S|^2+|S|-p_2p_4}v, s_1\hat{c}_6),$$
$$\Delta^{p_3}(a^{|S|-p_1-p_2}b^{|S|^2+|S|-p_2p_4}v, s_1\hat{c}_6) = (b^{|S|^2+2|S|-p_2p_4-p_1-p_2-p_3}v, s_2\hat{c}_7),$$
$$\Delta^{p_4}(b^{|S|^2+2|S|-p_2p_4-p_1-p_2-p_3}v, s_2\hat{c}_7) = (b^{|S|^2+2|S|-p_2p_4-p_1-p_2-p_3-p_4}v, s_2\hat{c}_5).$$

For the last equation, note that the length of the replaced factor is the same as of the original factor. That, is the number of steps is the same on both factors. Moreover, since there is no communication of the communication cell, during these steps the right part of the configuration develops as before. So, the configuration of the right part is finally $\hat{c}_5$.

Since the configurations after processing the original factor and its replacement are identical, the input word with replaced factor not belonging to L_{hi} is accepted as well. From the contradiction it follows that there is at least one communication while processing the factor $a^{|S|}b^{|S|^2+|S|}$.

Now we turn to the end of the computation. Assume that no communication takes place while processing the last $2|S|$ symbols from the last block, say these are symbols b. Let $w = ub^n$ with the last letter in u being an a. Then on processing the input suffix $b^{2|S|}$ the communication cell necessarily enters some state at least twice. Let $\Delta^{|u|+n-2|S|}(ub^n, s_0\hat{c}_0) = (b^{2|S|}, s\hat{c})$. Then there are $0 \le p_1$, $1 \le p_2$ with $p_1 + p_2 \le |S|$ such that $\Delta^{p_1}(b^{2|S|}, s\hat{c}) = (b^{2|S|-p_1}, s_1\hat{c}_1)$ and $\Delta^{p_2}(b^{2|S|-p_1}, s_1\hat{c}_1) = (b^{2|S|-p_1-p_2}, s_1\hat{c}_2)$. Since the input is accepted, the communication cell enters an accepting state in this state cycle of length p_2 (which cannot be left until the end of the computation) or before. This implies that input ub^{n-p_2} is accepted as well. From the contradiction it follows that there is at least one communication during the last $2|S|$ time steps.

Altogether we have seen that the communication cell of M necessarily communicates with its neighbor, or vice versa, during the first $|S|$ steps, during the last $2|S|$ steps, and on the factors $a^{|S|}b^{|S|^2+|S|}$ and $b^{|S|}a^{|S|^2+|S|}$ that include the block borders. Since there are $i - 1$ such block borders, in total, there are at least $i + 1$ communications for n long enough.   □

4   Undecidability Results for Realtime MC(*const*)-IAs

In this section, we first show that the question of emptiness is undecidable for realtime MC(*const*)-IAs by reduction of Hilbert's tenth problem which is known to be undecidable (see, e.g., [8,16]). The problem is to decide whether a given polynomial $p(x_1, \ldots, x_n)$ with integer coefficients has an integral root. That is, to decide whether there are integers $\alpha_1, \ldots, \alpha_n$ such that $p(\alpha_1, \ldots, \alpha_n) = 0$. A reduction of Hilbert's tenth problem to show the undecidability of emptiness for certain two-way counter machines has been used in [5] for the first time. A recent paper that provides a simulation of counter machines by linear-time one-way cellular automata is [1]. The basic idea of Ibarra in [5] is to define a language that encodes all possible values for $x_1, x_2, \ldots, x_n$ and evaluates the polynomial p for that values. Then, a two-counter machine is constructed accepting that language which is empty if and only if Hilbert's tenth problem has no solution in the integers. Such an approach we will in principal use here again. However, since we are concerned with IAs with constant communication and the input has to pass the communication cell to be processed by the IA, the definition of the language that evaluates the polynomial p is much more involved and will be presented in several steps.

As is remarked in [5], it is sufficient to restrict the variables $x_1, \ldots, x_n$ to take non-negative integers only. If $p(x_1, \ldots, x_n)$ contains a constant summand, then we may assume that it has a negative sign. Otherwise, we continue with $p(x_1, \ldots, x_n)$ multiplied with -1, whose constant summand now has a negative sign and which has the same integral roots as $p(x_1, \ldots, x_n)$. Such a polynomial has the following form:

$$p(x_1, \ldots, x_n) = t_1(x_1, \ldots, x_n) + \cdots + t_r(x_1, \ldots, x_n),$$

where each $t_j(x_1,\ldots,x_n)$, $1 \le j \le r$, is of the form

$$t_j(x_1,\ldots,x_n) = s_j x_1^{i_{j,1}} \cdots x_n^{i_{j,n}} \text{ with } s_j \neq 0 \text{ and } i_{j,1},\ldots,i_{j,n} \ge 0.$$

Since changing the sequence in which the summands appear does not change the polynomial, we additionally may assume that the summands are ordered according to their sign starting with summands having a positive sign. Moreover, we may assume that a constant term occurs at the end of the sequence only. Thus, $t_r = s_r$, if p contains s_r as constant. Finally, let $i_j = \sum_{k=1}^{n} i_{j,k}$.

Now, we consider a polynomial $p(x_1,\ldots,x_n)$ with integer coefficients that has the above form. Let t_j with $1 \le j \le r$ be a positive term. Then, we define language $L'(t_j)$ over the alphabet $\{b_0, b_1, \ldots, b_{i_j}\}$ as

$$L'(t_j) = \{ b_0^{s_j} b_1^{\alpha_1} \cdots b_{i_{j,1}}^{\alpha_1} b_{i_{j,1}+1}^{\alpha_2} \cdots b_{i_{j,1}+i_{j,2}}^{\alpha_2} \cdots b_{i_{j,1}+\cdots+i_{j,n-1}+1}^{\alpha_n} \cdots b_{i_j}^{\alpha_n} \mid$$

$$\alpha_1,\alpha_2,\ldots,\alpha_n \ge 0 \}.$$

Similarly, we define languages $L''(t_j)$ over $\{b_0, b_1, \ldots, b_{i_j}, c_1, \ldots, c_{i_j-1}, \$\}$ where words of $L'(t_j)$ are interleaved with c-blocks and a $\$$-block at the end, respectively, whose number of symbols is the product of the number of symbols of the two preceding blocks. So, the number of $\$$-symbols is an evaluation of $t_j(\alpha_1, \alpha_2, \ldots, \alpha_n)$.

$$L''(t_j) = \{ b_0^{s_j} b_1^{\alpha_1} c_1^{s_j \alpha_1} b_2^{\alpha_1} c_2^{s_j \alpha_1^2} b_3^{\alpha_1} c_3^{s_j \alpha_1^3} \cdots b_{i_{j,1}}^{\alpha_1} c_{i_{j,1}}^{s_j \alpha_1^{i_{j,1}}} b_{i_{j,1}+1}^{\alpha_2} c_{i_{j,1}+1}^{s_j \alpha_1^{i_{j,1}} \alpha_2} \cdots$$

$$b_{i_{j,1}+i_{j,2}}^{\alpha_2} c_{i_{j,1}+i_{j,2}}^{s_j \alpha_1^{i_{j,1}} \alpha_2^{i_{j,2}}} \cdots b_{i_j-1}^{\alpha_n} c_{i_j-1}^{s_j \alpha_1^{i_{j,1}} \alpha_2^{i_{j,2}} \cdots \alpha_n^{i_{j,n-1}}} b_{i_j}^{\alpha_n} \$^{s_j \alpha_1^{i_{j,1}} \alpha_2^{i_{j,2}} \cdots \alpha_n^{i_{j,n}}} \mid$$

$$\alpha_1,\alpha_2,\ldots,\alpha_n \ge 0 \}$$

Finally, $L(t_j)$ is defined over $\{b_0, b_1, \ldots, b_{i_j}, c_1, \ldots, c_{i_j-1}, d, \$\}$ as

$$L(t_j) = \{ wd^{7|w|} \mid w \in L''(t_j) \}.$$

If t_j with $1 \le j \le r$ is a negative and non-constant term, the definition of $L(t_j)$ is identical except for the fact that each symbol $\$$ is replaced by some symbol # and $b_0^{s_j}$ is replaced by $b_0^{|s_j|}$. If t_r is a constant term, we define $L(t_r) = \{\#^{|s_r|} d^{7|s_r|}\}$.

Example 5. Let $t_j(x_1, x_2, x_3, x_4) = 3x_1^2 x_2 x_4$. Then, $s_j = 3$, $i_{j,1} = 2$, $i_{j,2} = 1$, $i_{j,3} = 0$, $i_{j,4} = 1$, and $i_j = 4$, $L'(t_j) = \{ b_0^3 b_1^{\alpha_1} b_2^{\alpha_1} b_3^{\alpha_2} b_4^{\alpha_4} \mid \alpha_1, \alpha_2, \alpha_4 \ge 0 \}$ and $L''(t_j) = \{ b_0^3 b_1^{\alpha_1} c_1^{3\alpha_1} b_2^{\alpha_1} c_2^{3\alpha_1^2} b_3^{\alpha_2} c_3^{3\alpha_1^2 \alpha_2} b_4^{\alpha_4} \$^{3\alpha_1^2 \alpha_2 \alpha_4} \mid \alpha_1, \alpha_2, \alpha_4 \ge 0 \}$.

For example, to evaluate $t_j(2, 3, x_3, 2) = 3 \cdot 2^2 \cdot 3 \cdot 2 = 72$ we consider the word $b_0^3 b_1^2 c_1^6 b_2^2 c_2^{12} b_3^3 c_3^{36} b_4^2 \$^{72} d^{7 \cdot 138} \in L(t_j)$ and to evaluate $t_j(2, 1, x_3, 2) = 3 \cdot 2^2 \cdot 1 \cdot 2 = 24$ we consider the word $b_0^3 b_1^2 c_1^6 b_2^2 c_2^{12} b_3 c_3^{12} b_4^2 \$^{24} d^{7 \cdot 64} \in L(t_j)$. ∎

Let us start our construction with three lemmas that will be essential in the sequel.

Lemma 6. *An MC(const)-IA can effectively be constructed that shifts an input of the form $w \in a^+b^+c^+$ into its cells within $2|w|$ time steps.*

Lemma 7. *An MC(const)-IA can effectively be constructed that accepts every $w \in \{ a^n b^m c^n \mid n, m \geq 1 \}$ within $5|w|$ time steps.*

Lemma 8. *An MC(const)-IA can effectively be constructed that accepts every $w \in \{ a^n b^m c^{n \cdot m} \mid n, m \geq 1 \}$ within $7|w|$ time steps.*

Lemma 9. *The language $L(t_j)$ belongs to $\mathscr{L}_{rt}(MC(const)\text{-}IA)$ for each term t_j with $1 \leq j \leq r$.*

Proof. We describe the construction of a realtime MC($const$)-IA accepting $L(t_j)$, where t_j is a positive term. The construction for negative non-constant terms is identical except for exchanging \$ by #. If t_j is a negative constant term, $L(t_j)$ is a regular language and can be accepted by a realtime MC($const$)-IA using its communication cell only.

A realtime MC($const$)-IA for $L(t_j)$ basically computes four tasks.

1. Check the correct format of the input and check whether there are exactly s_j symbols b_0.
2. Check whether the length of every c-block is the product of the lengths of its two preceding blocks.
3. Check whether the number of d's is equal to seven times the length of the preceding input.
4. Check the equal number of symbols $b_{i_{j,1}+\cdots+i_{j,k-1}+1}, \ldots, b_{i_{j,1}+\cdots+i_{j,k}}$ for every $1 \leq k \leq n$.

The first task can be done by the communication cell which rejects the input in case of a wrong format or a wrong number of b_0's. For the second task we use the construction described in Lemma 8 multiple times. Having stored the complete input other than the d's in the array, the communication cell sends a signal with maximum speed to the right which starts in the b_0-block as well as in every c-block an instance of the construction given in Lemma 8 which checks whether the length of every c-block and of the last \$-block is the product of the lengths of its two preceding blocks. Whenever an error is encountered an error signal is sent to the left which rejects the input. Since there is only a finite number of c-blocks, we have to keep track of a finite number of instances of the construction of Lemma 8 which in addition are distributed over the array and at most two instances are overlapping. Since one instance can be realized by an MC($const$)-IA, we obtain that the finite number of such instances can be realized by an MC($const$)-IA as well. It has been shown in Lemma 8 that the total time for one instance is bounded by seven times the length of the input. Hence, we can conclude that, for some input wd^* with $w \in L''(t_j)$, time $7|w|$ is sufficient to accomplish the second task. This time is provided by the number of d's whose correct number is checked in the third task as follows. The communication cell sends a signal C with speed $1/6$ to the right which is reflected at the cell carrying the last \$ and is sent back to the left with maximum speed. The third task is successful if and only if C arrives again at the communication cell when the end-of-input symbol $\triangledown$ is read for the first time.

Finally, we have to accomplish the fourth task. Here, we have to check for every $1 \leq i \leq n$ whether each of a finite number, say n_i, of different b-symbols, say $\{b_{l_i}, b_{l_i+1}, \ldots, b_{l_i+n_i-1}\}$ is exactly α_i. This can basically be realized by implementing $n_i - 1$ instances of the construction given in Lemma 7. The first instance checks that the number of symbols b_{l_i} is equal to the number of symbols b_{l_i+1}, the second instance checks that the number of symbols b_{l_i+1} is equal to the number of symbols b_{l_i+2}, and so on. If all checks are positive for every $1 \leq i \leq n$, the fourth task is accomplished and requires at most $t = \sum_{i=1}^{n} 5\alpha_i(n_i - 1)$ time steps due to Lemma 7. Let $wd^{7|w|}$ with $w \in L''(t_j)$ be the input. Then, $\sum_{i=1}^{n} \alpha_i(n_i - 1)$ is bounded by $|w|$. Hence, $t \leq 5|w|$ and the number of d's provided gives enough time to accomplish the fourth task in realtime. Since n and all n_i for $1 \leq i \leq n$ are fixed numbers depending on t_j, we need only a finite number of instances of the construction given in Lemma 7 each of which can be realized by an MC($const$)-IA. Hence, we obtain that the fourth task can be realized by an MC($const$)-IA.

Altogether, by implementing the different tasks in parallel in different tracks we obtain a realtime MC($const$)-IA which accepts its input if and only if all four tasks have been accomplished successfully. $\square$

In Lemma 9 we have described how an evaluation of a term t_j can be simulated. Our next step is to simulate an evaluation of the given polynomial p. To this end, we have to put the evaluations of the single terms together. This will be done by concatenating certain regular languages around each language $L(t_j)$. Finally, the intersection of all these sets is considered which leads to an evaluation of all terms on the same input and thus to an evaluation of p.

Let us consider the following regular languages R_k depending on the sign of the term t_k. We set $R_k = b_0^* b_1^* c_1^* \cdots b_{i_k-1}^* c_{i_k-1}^* b_{i_k}^* \$^* d^*$ if $s_k > 0$, $R_k = \#^* d^*$ if t_k is the negative constant, and $R_k = b_0^* b_1^* c_1^* \cdots b_{i_k-1}^* c_{i_k-1}^* b_{i_k}^* \#^* d^*$ otherwise. Then, we define

$$\tilde{L}(t_j) = \{ a_1^{\alpha_1} \cdots a_n^{\alpha_n} w_1 w_2 \cdots w_r \mid \alpha_1, \ldots, \alpha_n \geq 0, w_i \in R_i, 1 \leq i \leq r, i \neq j,$$

$$w_j = w_j' d^{7|w_j'|}, \text{ and } w_j' = b_0^{s_j} b_1^{\alpha_1} c_1^{s_j \alpha_1} b_2^{\alpha_1} c_2^{s_j \alpha_1^2} b_3^{\alpha_1} c_3^{s_j \alpha_1^3} \cdots$$

$$b_{i_{j,1}}^{\alpha_1} c_{i_{j,1}}^{s_j \alpha_1^{i_{j,1}}} b_{i_{j,1}+1}^{\alpha_2} c_{i_{j,1}+1}^{s_j \alpha_1^{i_{j,1}} \alpha_2} \cdots b_{i_{j,1}+i_{j,2}}^{\alpha_2} c_{i_{j,1}+i_{j,2}}^{s_j \alpha_1^{i_{j,1}} \alpha_2^{i_{j,2}}} \cdots$$

$$b_{i_j-1}^{\alpha_n} c_{i_j-1}^{s_j \alpha_1^{i_{j,1}} \alpha_2^{i_{j,2}} \cdots \alpha_n^{i_{j,n-1}}} b_{i_j}^{\alpha_n} \$^{s_j \alpha_1^{i_{j,1}} \alpha_2^{i_{j,2}} \cdots \alpha_n^{i_{j,n}}} \}$$

and consider $\tilde{L}(p) = \bigcap_{j=1}^{r} \tilde{L}(t_j)$.

We can observe that each language $\tilde{L}(t_j)$ consists of the language $L(t_j)$, which evaluates $t_j(x_1, x_2, \ldots, x_n)$, and the concatenation of the regular sets $R_1 R_2 \cdots R_{j-1}$ to the left and $R_{j+1} \cdots R_r$ to the right. Additionally, each word of $\tilde{L}(t_j)$ starts with the substring $a_1^{\alpha_1} a_2^{\alpha_2} \cdots a_n^{\alpha_n}$ which determines the variables $x_1, x_2, \ldots, x_n$ to be evaluated. Since $\tilde{L}(p)$ is the intersection over all terms of the polynomial p, $\tilde{L}(p)$ simulates all those terms on the same input $\alpha_1, \alpha_2, \ldots, \alpha_n$. Thus, $\tilde{L}(p)$ denotes an evaluation of the given polynomial p whereby evaluations of positive and negative terms are encoded by \$-symbols and #-symbols, respectively.

Lemma 10. *The language $\tilde{L}(p)$ belongs to $\mathscr{L}_{rt}(MC(const)\text{-}IA)$.*

Finally, we set $L(p) = \{\, we^{4|w|} \mid w \in \tilde{L}(p), |w|_\$ = |w|_\# \,\}$ for some new alphabet symbol e.

Lemma 11. *The language $L(p)$ belongs to $\mathscr{L}_{rt}(MC(const)\text{-}IA)$.*

Proof. To accept $L(p)$ three checks are performed in parallel. First, it is verified that the prefix w belongs to $\tilde{L}(p)$ using the construction given in Lemma 10, which says in addition that the construction can be realized by an MC(*const*)-IA in realtime. Second, to check the correct number of e's the communication cell starts in the first time step a signal E_1 with speed $1/2$ to the right and starts another signal E_2 when the first e is read with maximum speed to the right. When both signals meet, some signal E_3 is sent back to the left with speed $1/3$ and has to arrive at the communication cell when the end-of-input symbol $\triangledown$ is read for the first time. Clearly, the second task can be done by a realtime MC(*const*)-IA. The third task is to check the equal number of \$- and #-symbols. The basic idea here is to implement again the construction given in Lemma 7. Since the terms of the given polynomial p are ordered with respect to their sign, we know that blocks of \$-symbols are followed by blocks of #-symbols. In between these blocks there are blocks of other alphabet symbols which have to be ignored. In a straightforward modification of the constructions given in Lemmas 6 and 7 we first enqueue a finite number of \$-blocks in one queue and then a finite number of #-blocks in a different queue. If both queues have an equal length, the third task is accomplished. According to Lemmas 6 and 7 we know that the third task can be done by an MC(*const*)-IA and needs at most $5|w|$ time steps. Due to the e-symbols provided the overall length of the input is $5|w|$ which implies that the third task can be accomplished in realtime.

Altogether, the input is accepted when all three task have successfully been done and we obtain that $L(p)$ is accepted by a realtime MC(*const*)-IA.   □

Now, all prerequisites are done and we obtain the following undecidability results for realtime MC(*const*)-IAs.

Theorem 12. *Emptiness is undecidable for realtime MC(const)-IAs.*

Proof. By applying Lemma 11 we can construct a realtime MC(*const*)-IA M accepting language $L(p)$ given a polynomial $p(x_1, \ldots, x_n)$. Moreover, the language $L(M) = L(p)$ is empty if and only if the polynomial $p(x_1, \ldots, x_n)$ has no solution in the non-negative integers Since Hilbert's tenth problem is undecidable, the emptiness problem for realtime MC(*const*)-IAs is undecidable as well.   □

Corollary 13. *The questions of finiteness, infiniteness, equivalence, and inclusion are undecidable for realtime MC(const)-IAs.*

References

1. Carton, O., Guillon, B., Reiter, F.: Counter machines and distributed automata a story about exchanging space and time. In: Baetens, J.M., Kutrib, M. (eds.) AUTOMATA 2018. LNCS, vol. 10875, pp. 13–28. Springer, Cham (2018). https:// doi.org/10.1007/978-3-319-92675-9_2
2. Chang, J.H., Ibarra, O.H., Palis, M.A.: Parallel parsing on a one-way array of finite-state machines. IEEE Trans. Comput. **C–36**, 64–75 (1987)
3. Cole, S.N.: Real-time computation by n-dimensional iterative arrays of finite-state machines. IEEE Trans. Comput. **C–18**(4), 349–365 (1969)
4. Fischer, P.C.: Generation of primes by a one-dimensional real-time iterative array. J. ACM **12**, 388–394 (1965)
5. Ibarra, O.H.: Reversal-bounded multicounter machines and their decision problems. J. ACM **25**(1), 116–133 (1978)
6. Ibarra, O.H., Palis, M.A.: Some results concerning linear iterative (systolic) arrays. J. Parallel Distrib. Comput. **2**, 182–218 (1985)
7. Ibarra, O.H., Palis, M.A.: Two-dimensional iterative arrays: characterizations and applications. Theor. Comput. Sci. **57**, 47–86 (1988)
8. Jones, J.P., Matijasevič, Y.V.: Proof of recursive unsolvability of Hilbert's tenth problem. Am. Math. Mon. **98**, 689–709 (1991)
9. Kutrib, M.: Cellular automata and language theory. In: Meyers, R.A. (ed.) Encyclopedia of Complexity and Systems Science, pp. 800–823. Springer, Berlin (2009). https://doi.org/10.1007/978-0-387-30440-3
10. Kutrib, M., Malcher, A.: Computations and decidability of iterative arrays with restricted communication. Parallel Process. Lett. **19**(2), 247–264 (2009)
11. Kutrib, M., Malcher, A.: On one-way one-bit $O(1)$-message cellular automata. Electr. Notes Theor. Comput. Sci. **252**, 77–91 (2009)
12. Kutrib, M., Malcher, A.: Cellular automata with sparse communication. Theor. Comput. Sci. **411**(38–39), 3516–3526 (2010)
13. Kutrib, M., Malcher, A.: One-way cellular automata, bounded languages, and minimal communication. J. Autom. Lang. Comb. **15**(1/2), 135–153 (2010)
14. Kutrib, M., Malcher, A.: Cellular automata with limited inter-cell bandwidth. Theor. Comput. Sci. **412**(30), 3917–3931 (2011)
15. Malcher, A.: Hierarchies and undecidability results for iterative arrays with sparse communication. In: Baetens, J.M., Kutrib, M. (eds.) AUTOMATA 2018. LNCS, vol. 10875, pp. 100–112. Springer, Cham (2018). https://doi.org/10.1007/978-3-319-92675-9_8
16. Matijasevič, Y.V.: On recursive unsolvability of Hilbert's tenth problem. In: Logic, Methodology and Philosophy of Science, IV (Proceedings of the Fourth International Congress, Bucharest, 1971), North-Holland, pp. 89–110 (1973)
17. Mazoyer, J., Terrier, V.: Signals in one-dimensional cellular automata. Theor. Comput. Sci. **217**(1), 53–80 (1999)
18. Umeo, H., Kamikawa, N.: A design of real-time non-regular sequence generation algorithms and their implementations on cellular automata with 1-bit inter-cell communications. Fundam. Inf. **52**, 257–275 (2002)
19. Umeo, H., Kamikawa, N.: Real-time generation of primes by a 1-bit-communication cellular automaton. Fundam. Inf. **58**, 421–435 (2003)
20. Worsch, T.: Linear time language recognition on cellular automata with restricted communication. In: Gonnet, G.H., Viola, A. (eds.) LATIN 2000. LNCS, vol. 1776, pp. 417–426. Springer, Heidelberg (2000). https://doi.org/10.1007/10719839_41

Bounding the Minimal Number
of Generators of Groups and Monoids
# of Cellular Automata

Alonso Castillo-Ramirez$^{(\boxtimes)}$ (iD) and Miguel Sanchez-Alvarez

Department of Mathematics, University Center of Exact Sciences and Engineering,
University of Guadalajara, Guadalajara, Mexico
alonso.castillor@academicos.udg.mx, miguel_201288@hotmail.com

Abstract. For a group G and a finite set A, denote by $\mathrm{CA}(G;A)$ the
monoid of all cellular automata over A^G and by $\mathrm{ICA}(G;A)$ its group of
units. We study the minimal cardinality of a generating set, known as
the *rank*, of $\mathrm{ICA}(G;A)$. In the first part, when G is a finite group, we give
upper bounds for the rank in terms of the number of conjugacy classes
of subgroups of G. The case when G is a finite cyclic group has been
studied before, so here we focus on the cases when G is a finite dihedral
group or a finite Dedekind group. In the second part, we find a basic
lower bound for the rank of $\mathrm{ICA}(G;A)$ when G is a finite group, and we
apply this to show that, for any infinite abelian group H, the monoid
$\mathrm{CA}(H;A)$ is not finitely generated. The same is true for various kinds of
infinite groups, so we ask if there exists an infinite group H such that
$\mathrm{CA}(H;A)$ is finitely generated.

Keywords: Monoid of cellular automata ·
Invertible cellular automata · Minimal number of generators

1 Introduction

The theory of cellular automata (CA) has important connections with many
areas of mathematics, such as group theory, topology, symbolic dynamics, coding theory, and cryptography. Recently, in [5–7], links with semigroup theory
have been explored, and, in particular, questions have been considered on the
structure of the monoid of all CA and the group of all invertible CA over a given
configuration space. The goal of this paper is to bound the minimal number of
generators, known in semigroup theory as the *rank*, of groups of invertible CA.

Let G be a group and A a finite set. Denote by A^G the *configuration space*,
i.e. the set of all functions of the form $x : G \to A$. The *shift action* of G on A^G
is defined by $g \cdot x(h) = x(g^{-1}h)$, for all $x \in A^G$, $g, h \in G$. We endow A^G with
the *prodiscrete topology*, which is the product topology of the discrete topology
on A. A *cellular automaton* over A^G is a transformation $\tau : A^G \to A^G$ such that
there is a finite subset $S \subseteq G$ and a function $\mu : A^S \to A$ satisfying

$$\tau(x)(g) = \mu((g^{-1} \cdot x)|_S), \forall x \in A^G, g \in G,$$

© IFIP International Federation for Information Processing 2019
Published by Springer Nature Switzerland AG 2019
A. Castillo-Ramirez and P. P. B. de Oliveira (Eds.): AUTOMATA 2019, LNCS 11525, pp. 48–61, 2019.
https://doi.org/10.1007/978-3-030-20981-0_4

where $|_S$ denotes the restriction to S of a configuration in A^G.

Curtis-Hedlund Theorem [8, Theorem 1.8.1] establishes that a function $\tau :$ $A^G \to A^G$ is a cellular automaton if and only if it is continuous in the prodiscrete topology of A^G and commutes with the shift action (i.e. $\tau(g \cdot x) = g \cdot \tau(x)$, for all $x \in A^G$, $g \in G$). By [8, Corollary 1.4.11], the composition of two cellular automata over A^G is a cellular automaton over A^G. This implies that, equipped with composition of functions, the set $\mathrm{CA}(G; A)$ of all cellular automata over A^G is a monoid. The *group of units* (i.e. group of invertible elements) of $\mathrm{CA}(G; A)$ is denoted by $\mathrm{ICA}(G; A)$. When $|A| \geq 2$ and $G = \mathbb{Z}$, many interesting properties are known for $\mathrm{ICA}(\mathbb{Z}; A)$: for example, every finite group, as well as the free group on a countable number of generators, may be embedded in $\mathrm{ICA}(\mathbb{Z}; A)$ (see [2]). However, despite of several efforts, most of the algebraic properties of $\mathrm{CA}(G; A)$ and $\mathrm{ICA}(G; A)$ still remain unknown.

Given a subset T of a monoid M, the *submonoid generated* by T, denoted by $\langle T \rangle$, is the smallest submonoid of M that contains T; this is equivalent as defining $\langle T \rangle := \{t_1 t_2 \dots t_k \in M : t_i \in T, \forall i, k \geq 0\}$. We say that T is a *generating set of M* if $M = \langle T \rangle$. The monoid M is said to be *finitely generated* if it has a finite generating set. The *rank* of M is the minimal cardinality of a generating set:

$$\mathrm{Rank}(M) := \min\{|T| : M = \langle T \rangle\}.$$

The question of finding the rank of a monoid is important in semigroup theory; it has been answered for several kinds of transformation monoids and Rees matrix semigroups (e.g., see [1, 10, 11]). For the case of monoids of cellular automata over finite groups, the question has been addressed in [6, 7]; in particular, the rank of $\mathrm{ICA}(G; A)$ when G is a finite cyclic group has been examined in detail in [5].

For any subset U of a monoid M, the *relative rank* of U in M is

$$\mathrm{Rank}(M : U) = \min\{|W| : M = \langle U \cup W \rangle\}.$$

When U is the group of units of M, we have the basic identity

$$\mathrm{Rank}(M) = \mathrm{Rank}(M : U) + \mathrm{Rank}(U), \tag{1}$$

which follows as the group U can only be generated by subsets of itself. The relative rank of $\mathrm{ICA}(G; A)$ in $\mathrm{CA}(G; A)$ has been established in [7, Theorem 7] for finite *Dedekind groups* (i.e. groups in which all subgroups are normal).

In this paper, we study the rank of $\mathrm{ICA}(G; A)$ when G is a finite group. In Sect. 2, we introduce notation and review some basic facts, including the structure theorem for $\mathrm{ICA}(G; A)$ obtained in [7]. In Sect. 3, we give upper bounds for the rank of $\mathrm{ICA}(G; A)$, examining in detail the cases when G is a finite dihedral group or a finite Dedekind group, but also obtaining some results for a general finite group. In Sect. 4, we show that, when G is finite, the rank of $\mathrm{ICA}(G; A)$ is at least the number of conjugacy classes of subgroups of G. As an application, we use this to provide a simple proof that the monoid $\mathrm{CA}(G; A)$ is not finitely generated whenever G is an infinite abelian group. This result implies that $CA(G; A)$ is not finitely generated for various classes of infinite groups, such as free groups and the infinite dihedral group. Thus, we ask if there exists an infinite group G such that the monoid $CA(G; A)$ is finitely generated.

2   Basic Results

We assume the reader has certain familiarity with basic concepts of group theory.

Let G be a group and A a finite set. The *stabiliser* and *G-orbit* of a configuration $x \in A^G$ are defined, respectively, by

$$G_x := \{g \in G : g \cdot x = x\} \text{ and } Gx := \{g \cdot x : g \in G\}.$$

Stabilisers are subgroups of G, while the set of G-orbits forms a partition of A^G.

Two subgroups H_1 and H_2 of G are *conjugate* in G if there exists $g \in G$ such that $g^{-1}H_1 g = H_2$. This defines an equivalence relation on the subgroups of G. Denote by $[H]$ the conjugacy class of $H \leq G$. A subgroup $H \leq G$ is *normal* if $[H] = \{H\}$ (i.e. $g^{-1}Hg = H$ for all $g \in G$). Let $N_G(H) := \{g \in G : H = g^{-1}Hg\} \leq G$ be the *normaliser of H in G*. Note that H is always a normal subgroup of $N_G(H)$. Denote by $r(G)$ the total number of conjugacy classes of subgroups of G, and by $r_i(G)$ the number of conjugacy classes $[H]$ such that H has index i in G:

$$r(G) := |\{[H] : H \leq G\}|,$$
$$r_i(G) := |\{[H] : H \leq G, \ [G : H] = i\}|.$$

For any $H \leq G$, denote

$$\alpha_{[H]}(G; A) := |\{Gx \subseteq A^G : [G_x] = [H]\}|.$$

This number may be calculated using the Mobius function of the subgroup lattice of G, as shown in [7, Sect. 4].

For any integer $\alpha \geq 1$, let S_α be the symmetric group of degree α. The *wreath product* of a group C by S_α is the set

$$C \wr S_\alpha := \{(v; \phi) : v \in C^\alpha, \phi \in S_\alpha\}$$

equipped with the operation $(v; \phi) \cdot (w; \psi) = (v \cdot w^\phi; \phi\psi)$, for any $v, w \in C^\alpha, \phi, \psi \in S_\alpha$, where ϕ acts on w by permuting its coordinates:

$$w^\phi = (w_1, w_2, \ldots, w_\alpha)^\phi := (w_{\phi(1)}, w_{\phi(2)}, \ldots, w_{\phi(\alpha)}).$$

In fact, as may be seen from the above definitions, $C \wr S_\alpha$ is equal to the external semidirect product $C^\alpha \rtimes_\varphi S_\alpha$, where $\varphi : S_\alpha \to \mathrm{Aut}(C^\alpha)$ is the action of S_α of permuting the coordinates of C^α. For a more detailed description of the wreath product see [1].

Theorem 1 ([7]). *Let G be a finite group and A a finite set of size $q \geq 2$. Let $[H_1], \ldots, [H_r]$ be the list of all different conjugacy classes of subgroups of G. Let $\alpha_i := \alpha_{[H_i]}(G; A)$. Then,*

$$ICA(G; A) \cong \prod_{i=1}^{r} ((N_G(H_i)/H_i) \wr S_{\alpha_i}).$$

3   Upper Bounds for Ranks

The Rank function on monoids does not behave well when taking submonoids or subgroups: in other words, if N is a submonoid of M, there may be no relation between $\mathrm{Rank}(N)$ and $\mathrm{Rank}(M)$. For example, if $M = S_n$ is the symmetric group of degree $n \geq 3$ and N is a subgroup of S_n generated by $\lfloor \frac{n}{2} \rfloor$ commuting transpositions, then $\mathrm{Rank}(S_n) = 2$, as S_n may be generated by a transposition and an n-cycle, but $\mathrm{Rank}(N) = \lfloor \frac{n}{2} \rfloor$. It is even possible that M is finitely generated but N is not finitely generated (such as the case of the free group on two symbols and its commutator subgroup). However, the following lemma gives us some elementary tools to bound the rank in some cases.

Lemma 1. *Let G and H be a groups, and let N be a normal subgroup of G. Then:*

1. $\mathrm{Rank}(G/N) \leq \mathrm{Rank}(G)$.
2. $\mathrm{Rank}(G \times H) \leq \mathrm{Rank}(G) + \mathrm{Rank}(H)$.
3. $\mathrm{Rank}(G \wr S_\alpha) \leq \mathrm{Rank}(G) + \mathrm{Rank}(S_\alpha)$, *for any $\alpha \geq 1$.*
4. $\mathrm{Rank}(\mathbb{Z}_d \wr S_\alpha) = 2$, *for any $d, \alpha \geq 2$.* .

Proof. Parts 1 and 2 are straightforward. For parts 3 and 4, see [7, Corollary 5] and [5, Lemma 5], respectively.   □

We shall use Lemma 1 together with Theorem 1 in order to find upper bounds for $\mathrm{ICA}(G; A)$. Because of part 3 in Lemma 1, it is now relevant to determine some values of the α_i's that appear in Theorem 1.

Lemma 2. *Let G be a finite group and A a finite set of size $q \geq 2$. Let H be a subgroup of G.*

1. $\alpha_{[G]}(G; A) = q$.
2. $\alpha_{[H]}(G; A) = 1$ *if and only if $[G : H] = 2$ and $q = 2$.*
3. *If $q \geq 3$, then $\alpha_{[H]}(G; A) \geq 3$.*

Proof. Parts 1 and 2 correspond to Remark 1 and Lemma 5 in [7], respectively. For part 2, Suppose that $q \geq 3$ and $\{0, 1, 2\} \subseteq A$. Define configurations $z_1, z_2, z_3 \in A^G$ as follows,

$$z_1(g) = \begin{cases} 1 & \text{if } g \in H \\ 0 & \text{if } g \notin H, \end{cases} \quad z_2(g) = \begin{cases} 2 & \text{if } g \in H \\ 0 & \text{if } g \notin H, \end{cases} \quad z_3(g) = \begin{cases} 1 & \text{if } g \in H \\ 2 & \text{if } g \notin H, \end{cases}$$

All three configurations are in different orbits and $G_{z_i} = H$, for $i = 1, 2, 3$. Hence $\alpha_{[H]}(G; A) \geq 3$.   □

Although we shall not use explicitly part 3 of the previous lemma, the result is interesting as it shows that, for $q \geq 3$, our upper bounds cannot be refined by a more careful examination of the values of the $\alpha_i's$, as, for all $\alpha \geq 3$, we have $\mathrm{Rank}(S_\alpha) = 2$.

3.1 Dihedral Groups

In this section we investigate the rank of $\mathrm{ICA}(D_{2n}; A)$, where D_{2n} is the dihedral group of order $2n$, with $n \geq 1$, and A is a finite set of size $q \geq 2$. We shall use the following standard presentation of D_{2n}:

$$D_{2n} = \langle \rho, s \mid \rho^n = s^2 = s\rho s\rho = \mathrm{id} \rangle.$$

Lemma 3. *For any $n \geq 2$ and $\alpha \geq 2$, $\mathrm{Rank}(D_{2n} \wr S_\alpha) \leq 3$.*

Proof. By [5, Lemma 5], we know that $\langle \rho \rangle \wr S_\alpha \cong \mathbb{Z}_n \wr S_\alpha$ may be generated by two elements. Hence, by adding $((s, \mathrm{id}, \ldots, \mathrm{id}); \mathrm{id})$ we may generate the whole $D_{2n} \wr S_\alpha$ with three elements. $\quad\square$

Given a subgroup $H \leq D_{2n}$, we shall now analyze the quotient group $N_G(H)/H$.

Lemma 4. *Let $G = D_{2n}$ and let $H \leq D_{2n}$ be a subgroup of odd index m. Then H is self-normalizing, i.e. $N_G(H) = H$.*

Proof. By [9, Theorem 3.3], all subgroups of D_{2n} with index m are conjugate to each other. By [9, Corollary 3.2], there are m subgroups of D_{2n} of index m, so $|[H]| = m = [D_{2n} : H]$. On the other hand, by the Orbit-Stabilizer Theorem applied to the conjugation action of D_{2n} on its subgroups we have $|[H]| = [D_{2n} : N_G(H)]$. Therefore, $[D_{2n} : H] = [D_{2n} : N_G(H)]$ and $N_G(H) = H$. $\quad\square$

Lemma 5. *Let $G = D_{2n}$ and let $H \leq D_{2n}$ be a proper subgroup of even index m. Then H is normal in D_{2n} and $D_{2n}/H \cong D_m$, except when n is even, $m \mid n$, and $[H] = [\langle \rho^m, s \rangle]$ or $[H] = [\langle \rho^m, \rho s \rangle]$, in which case $N_G(H)/H \cong \mathbb{Z}_2$.*

Proof. We shall use Corollary 3.2 and Theorem 3.3 in [9]. There are two cases to consider:

1. Suppose n is odd. Then, D_{2n} has a unique subgroup of index m, so $1 = |[H]| = [D_{2n} : N_G(H)]$. This implies that $D_{2n} = N_G(H)$, so H is normal in D_{2n}.
2. Suppose that n is even. If $m \nmid n$, D_{2n} has a unique subgroup of index m, so H is normal by the same argument as in the previous case. If $m \mid n$, then D_{2n} has $m + 1$ subgroups with index m partitioned into 3 conjugacy classes $[\langle \rho^{m/2} \rangle]$, $[\langle \rho^m, s \rangle]$ and $[\langle \rho^m, rs \rangle]$ of sizes 1, $\frac{m}{2}$ and $\frac{m}{2}$, respectively. If $|[H]| = 1$, again H is normal. If $|[H]| = \frac{m}{2}$, then $[D_{2n} : N_G(H)] = \frac{1}{2}[D_{2n} : H]$. Hence, $[N_G(H) : H] = 2$, and $N_G(H)/H \cong \mathbb{Z}_2$.

The fact that $D_{2n}/H \cong D_m$ whenever H is normal follows by [9, Theorem 2.3]. $\quad\square$

Let $d(n)$ be number of divisors of n, including 1 and n itself. Let $d_-(n)$ and $d_+(n)$ be the number of odd and even divisors of n, respectively.

When n is odd, D_{2n} has exactly 1 conjugacy class of subgroups of index m, for every $m \mid 2n$; hence, if n is odd, $r(D_{2n}) = d(2n)$. When n is even, D_{2n} has exactly 1 conjugacy class of index m, when $m \mid 2n$ is odd or $m \nmid n$, and exactly 3 conjugacy classes when $m \mid 2n$ is even and $m \mid n$; hence, if n is even, $r(D_{2n}) = d(2n) + 2d_+(n)$.

Theorem 2. *Let $n \geq 3$ be an integer and A a finite set of size at least 2.*

$$\text{Rank}(\text{ICA}(D_{2n}; A)) \leq \begin{cases} 2d_-(2n) + 3d_+(2n) - 3 & \text{if } n \text{ is odd and } q = 2, \\ 2d_-(2n) + 3d_+(2n) - 1 & \text{if } n \text{ is odd and } q \geq 3, \\ 2d_-(2n) + 3d_+(2n) + 2d_+(n) - 3 & \text{if } n \text{ is even and } q = 2, \\ 2d_-(2n) + 3d_+(2n) + 4d_+(n) - 1 & \text{if } n \text{ is even and } q \geq 3. \end{cases}$$

Proof. Let $[H_1], \ldots, [H_r]$ be the conjugacy classes of subgroups of D_{2n}. Let $m_i = [G : H_i]$, with $m_{r-1} = 2$ and $m_r = 1$ (i.e. $H_r = D_{2n}$). Define $\alpha_i := \alpha_{[H_i]}(G; A)$. Let n be odd. Then, by Theorem 1, and Lemmas 4 and 5,

$$\text{ICA}(D_{2n}; A) \cong \prod_{m_i \mid 2n \text{ odd}} S_{\alpha_i} \times \prod_{2 < m_i \mid 2n \text{ even}} (D_{m_i} \wr S_{\alpha_i}) \times (\mathbb{Z}_2 \wr S_{\alpha_{r-1}})$$

By Lemmas 1 and 3,

$$\text{Rank}(\text{ICA}(D_{2n}; A)) \leq \sum_{m_i \mid 2n \text{ odd}} \text{Rank}(S_{\alpha_i}) + \sum_{2 < m_i \mid 2n \text{ even}} \text{Rank}(D_{m_i} \wr S_{\alpha_i}) + 2$$

$$\leq 2d_-(2n) + 3(d_+(2n) - 1) + 2$$

$$= 2d_-(2n) + 3d_+(2n) - 1.$$

When $q = 2$, Lemma 2 shows that $\alpha_r = 2$ and $\alpha_{r-1} = 1$, so $S_{\alpha_r} \cong S_2$ and $\mathbb{Z}_2 \wr S_{\alpha_{r-1}} \cong \mathbb{Z}_2$. Therefore,

$$\text{Rank}(\text{ICA}(D_{2n}; A)) \leq 1 + \sum_{1 < m_i \mid 2n \text{ odd}} 2 + \sum_{2 < m_i \mid 2n \text{ even}} 3 + 1$$

$$\leq 2(d_-(2n) - 1) + 3(d_+(2n) - 1) + 2$$

$$= 2d_-(2n) + 3d_+(2n) - 3.$$

Now let n be even. Then,

$$\text{ICA}(D_{2n}; A) \cong \prod_{m_i \mid 2n \text{ odd}} S_{\alpha_i} \times \prod_{2 < m_i \mid 2n \text{ even}} (D_{m_i} \wr S_{\alpha_i})$$

$$\times (\mathbb{Z}_2 \wr S_{\alpha_{r-1}}) \times \prod_{m_i \mid n \text{ even}} ((\mathbb{Z}_2 \wr S_{\alpha_i}) \times (\mathbb{Z}_2 \wr S_{\alpha_i}))$$

Hence,

$$\text{Rank}(\text{ICA}(D_{2n}; A)) \le \sum_{m_i | 2n \text{ odd}} 2 + \sum_{2 < m_i | 2n \text{ even}} 3 + 2 + \sum_{m_i | n \text{ even}} 4$$

$$= 2\text{d}_-(2n) + 3(\text{d}_+(2n) - 1) + 2 + 4\text{d}_+(n)$$

$$= 2\text{d}_-(2n) + 3\text{d}_+(2n) + 4\text{d}_+(n) - 1.$$

When $q = 2$, by Lemma 2 we have $S_{\alpha_r} \cong S_2$, $\mathbb{Z}_2 \wr S_{\alpha_{r-1}} \cong \mathbb{Z}_2$ and $\mathbb{Z}_2 \wr S_{\alpha_i} \cong \mathbb{Z}_2$, so

$$\text{Rank}(\text{ICA}(D_{2n}; A)) \le 1 + \sum_{1 < m_i | 2n \text{ odd}} 2 + \sum_{2 < m_i | 2n \text{ even}} 3 + 1 + \sum_{m_i | n \text{ even}} 2$$

$$= 2(\text{d}_-(2n) - 1) + 3(\text{d}_+(2n) - 1) + 2\text{d}_+(n) + 2$$

$$= 2\text{d}_-(2n) + 3\text{d}_+(2n) + 2\text{d}_+(n) - 3.$$

$\square$

Example 1. Let A be a finite set of size $q \ge 2$. By the previous theorem,

$$\text{Rank}(\text{ICA}(D_6; A)) \le \begin{cases} 2d_-(6) + 3d_+(6) - 3 = 2 \cdot 2 + 3 \cdot 2 - 3 = 7 & \text{if } q = 2, \\ 2d_-(6) + 3d_+(6) - 1 = 2 \cdot 2 + 3 \cdot 2 - 1 = 9 & \text{if } q \ge 2. \end{cases}$$

On the other hand,

$$\text{Rank}(\text{ICA}(D_8; A)) \le \begin{cases} 2d_-(8) + 3d_+(8) + 2d_+(4) - 3 = 12 & \text{if } q = 2, \\ 2d_-(8) + 3d_+(8) + 4d_+(4) - 1 = 18 & \text{if } q \ge 2. \end{cases}$$

3.2   Other Finite Groups

Recall that $r(G)$ denotes the total number of conjugacy classes of subgroups of G and $r_i(G)$ the number of conjugacy classes $[H]$ such that H has index i in G. The following results are an improvement of [7, Corollary 5].

Theorem 3. *Let G be a finite Dedekind group and A a finite set of size $q \ge 2$. Let $r := r(G)$ and $r_i := r_i(G)$. Let $p_1, \ldots, p_s$ be the prime divisors of $|G|$ and define $r_P := \sum_{i=1}^{s} r_{p_i}$. Then,*

$$\text{Rank}(\text{ICA}(G; A)) \le \begin{cases} (r - r_P - 1)\text{Rank}(G) + 2r - r_2 - 1, & \text{if } q = 2, \\ (r - r_P - 1)\text{Rank}(G) + 2r, & \text{if } q \ge 3. \end{cases}$$

Proof. Let $H_1, H_2, \ldots, H_r$ be the list of different subgroups of G with $H_r = G$. If H_i is a subgroup of index p_k, then $(G/H_i) \wr S_{\alpha_i} \cong \mathbb{Z}_{p_k} \wr S_{\alpha_i}$ is a group with rank 2, by Lemma 1. Thus, by Theorem 1 we have:

$$\mathrm{Rank}(\mathrm{ICA}(G; A)) \leq \sum_{i=1}^{r-1} \mathrm{Rank}((G/H_i) \wr S_{\alpha_i}) + \mathrm{Rank}(S_q)$$

$$\leq \sum_{[G:H_i]=p_k} 2 + \sum_{[G:H_i] \neq p_k} (\mathrm{Rank}(G) + 2) + 2$$

$$= 2r_P + (r - r_P - 1)(\mathrm{Rank}(G) + 2) + 2$$

$$= (r - r_P - 1)\mathrm{Rank}(G) + 2r.$$

If $q = 2$, we may improve this bound by using Lemma 2:

$$\mathrm{Rank}(\mathrm{ICA}(G; A)) \leq \sum_{[G:H_i]=2} \mathrm{Rank}((G/H_i) \wr S_1) + \sum_{[G:H_i]=p_k \neq 2} \mathrm{Rank}((G/H_i) \wr S_{\alpha_i})$$

$$+ \sum_{1 \neq [G:H_i] \neq p_k} \mathrm{Rank}((G/H_i) \wr S_{\alpha_i}) + \mathrm{Rank}(S_2)$$

$$= r_2 + 2(r_P - r_2) + (r - r_P - 1)(\mathrm{Rank}(G) + 2) + 1$$

$$= (r - r_P - 1)\mathrm{Rank}(G) + 2r - r_2 - 1.$$

□

Example 2. The smallest example of a nonabelian Dedekind group is the quaternion group

$$Q_8 = \langle x, y \mid x^4 = x^2 y^{-2} = y^{-1} x y x = \mathrm{id} \rangle,$$

which has order 8. It is generated by two elements, and it is noncyclic, so $\mathrm{Rank}(Q_8) = 2$. Moreover, $r = r(Q_8) = 6$ and, as 2 is the only prime divisor of 8, we have $r_P = r_2 = 3$. Therefore,

$$\mathrm{Rank}(\mathrm{ICA}(Q_8; A)) \leq \begin{cases} (6 - 3 - 1) \cdot 2 + 2 \cdot 6 - 3 - 1 = 12, & \text{if } q = 2, \\ (6 - 3 - 1) \cdot 2 + 2 \cdot 6 = 16, & \text{if } q \geq 3. \end{cases}$$

Corollary 1. *Let G be a finite Dedekind group and A a finite set of size $q \geq 2$. With the notation of Theorem 3,*

$$\mathrm{Rank}(\mathrm{CA}(G; A)) \leq \begin{cases} (r - r_P - 1)\mathrm{Rank}(G) + \frac{1}{2}r(r + 5) - 2r_2 - 1, & \text{if } q = 2 \\ (r - r_P - 1)\mathrm{Rank}(G) + \frac{1}{2}r(r + 5), & \text{otherwise.} \end{cases}$$

Proof. The result follows by Theorem 3, identity (1) and the basic upper bound for the relative rank that follows from [7, Theorem 7]:

$$\mathrm{Rank}(\mathrm{CA}(G; A) : \mathrm{ICA}(G; A)) \leq \begin{cases} \binom{r}{2} + r - r_2 & \text{if } q = 2 \\ \binom{r}{2} + r, & \text{otherwise.} \end{cases}$$

□

We focus now when G is not necessarily a Dedekind group.

Lemma 6. *Let G be a finite group and H a subgroup of G of prime index p. Let A be a finite set of size $q \geq 2$ and $\alpha := \alpha_{[H]}(G; A)$. Then*

$$\mathrm{Rank}\left((N_G(H)/H) \wr S_\alpha\right) \leq \begin{cases} 1 & \text{if } p = 2 \text{ and } q = 2 \\ 2 & \text{otherwise.} \end{cases}$$

Proof. By Lagrange's theorem, $N_G(H) = H$ or $N_G(H) = G$. Hence, in order to find an upper bound for the above rank, we assume that H is normal in G. As the index is prime, $G/H \cong \mathbb{Z}_p$. If $p = 2$ and $q = 2$, Lemma 2 shows that $\alpha = 1$, so $\mathrm{Rank}(\mathbb{Z}_2 \wr S_1) = 1$. For the rest of the cases we have that $\mathrm{Rank}(\mathbb{Z}_p \wr S_\alpha) = 2$, by Lemma 1. □

The *length* of G (see [3, Sect. 1.15]) is the length $\ell := \ell(G)$ of the longest chain of proper subgroups

$$1 = G_0 < G_1 < \cdots < G_\ell = G.$$

The lengths of the symmetric groups are known by [4]: $\ell(S_n) = \lceil 3n/2 \rceil - b(n) - 1$, where $b(n)$ is the numbers of ones in the base 2 expansion of n. As, $\ell(G) = \ell(N) + \ell(G/N)$ for any normal subgroup N of G, the length of a finite group is equal to the sum of the lengths of its compositions factors; hence, the question of calculating the length of all finite groups is reduced to calculating the length of all finite simple groups. Moreover, $\ell(G) \leq \log_2(|G|)$ (see [4, Lemma 2.2]).

Lemma 7. *Let G be a finite group and H a subgroup of G. Let A be a finite set of size $q \geq 2$ and $\alpha := \alpha_{[H]}(G; A)$. Then,*

$$\mathrm{Rank}\left((N_G(H)/H) \wr S_\alpha\right) \leq \ell(G) + 2$$

Proof. By Lemma 1, $\mathrm{Rank}\left((N_G(H)/H) \wr S_\alpha\right) \leq \mathrm{Rank}(N_G(H)) + 2$. Observe that $\mathrm{Rank}(G) \leq \ell(G)$, as the set $\{g_i \in G : g_i \in G_i - G_{i-1},\ i = 1, \ldots, \ell\}$ (with G_i as the above chain of proper subgroups) generates G. Moreover, it is clear that $\ell(K) \leq \ell(G)$ for every subgroup $K \leq G$, so the result follows by letting $K = N_G(H)$. □

Theorem 4. *Let G be a finite group of size n, $r := r(G)$, and A a finite set of size $q \geq 2$. Let r_i be the number of conjugacy classes of subgroups of G of index i. Let $p_1, \ldots, p_s$ be the prime divisors of $|G|$ and let $r_P = \sum_{i=1}^{s} r_i$. Then:*

$$\mathrm{Rank}(\mathrm{ICA}(G; A)) \leq \begin{cases} (r - r_P - 1)\ell(G) + 2r - r_2 - 1 & \text{if } q = 2, \\ (r - r_P - 1)\ell(G) + 2r & \text{if } q \geq 3. \end{cases}$$

Proof. Let $H_1, H_2, \ldots, H_r$ be the list of different subgroups of G with $H_r = G$. By Theorem 1 and Lemmas 1, 6, 7,

$$\text{Rank}(\text{ICA}(G; A)) \leq \sum_{i=1}^{r-1} \text{Rank}\left((N_G(H_i)/H_i) \wr S_{\alpha_i}\right) + \text{Rank}(S_q)$$

$$\leq \sum_{[G:H_i]=p_k} 2 + \sum_{1 \neq [G:H_i] \neq p_k} (\ell(G) + 2) + 2$$

$$= 2r_P + (r - r_P - 1)(\ell(G) + 2) + 2$$

$$= (r - r_P - 1)\ell(G) + 2r.$$

When $q = 2$, we may improve this bound as follows:

$$\text{Rank}(\text{ICA}(G; A)) \leq \sum_{[G:H_i]=2} 1 + \sum_{[G:H_i]=p_k \neq 2} 2 + \sum_{1 < [G:H_i] \neq p_k} (\ell(G) + 2) + 1$$

$$= r_2 + 2(r_P - r_2) + (r - r_P - 1)(\ell(G) + 2) + 1$$

$$= (r - r_P - 1)\ell(G) + 2r - r_2 - 1.$$

$\square$

If G is a subgroup of S_n, we may find a good upper bound for Rank $(\text{ICA}(G; A))$ in terms of n by using a theorem of McIver and Neumann.

Proposition 1. *Suppose that $G \leq S_n$, for some $n > 3$. Let $r := r(G)$. Then*

$$\text{Rank}(\text{ICA}(G; A)) \leq \begin{cases} (r-1)\left\lfloor \frac{n}{2} \right\rfloor + 2r - r_2 - 1 & \text{if } q = 2, \\ (r-1)\left\lfloor \frac{n}{2} \right\rfloor + 2r & \text{if } q \geq 3. \end{cases}$$

Proof. By [12], for every $n > 3$ and every $K \leq S_n$, $\text{Rank}(K) \leq \lfloor \frac{n}{2} \rfloor$. The rest of the proof is analogous to the previous one. $\square$

Example 3. Consider the symmetric group S_4. In this case it is known that $r = r(S_4) = 11$ and $r_2 = 1$ (as A_4 is its only subgroup of index 2). Therefore,

$$\text{Rank}(\text{ICA}(S_4; A)) \leq \begin{cases} (11-1)\frac{4}{2} + 2 \cdot 11 - 1 - 1 = 40 & \text{if } q = 2, \\ (11-1)\frac{4}{2} + 2 \cdot 11 = 42 & \text{if } q \geq 3. \end{cases}$$

For sake of comparison, the group $\text{ICA}(S_4; \{0, 1\})$ has order $2^{2^{24}}$.

4   Lower Bounds on Ranks

4.1   Finite Groups

Proposition 2. *Let G be a finite group and A a finite set of size $q \geq 2$. Then*

$$\text{Rank}(\text{ICA}(G; A) \geq \begin{cases} r(G) - r_2(G) & \text{if } q = 2, \\ r(G) & \text{otherwise.} \end{cases}$$

Proof. Let $[H_1], [H_2], \ldots, [H_r]$ be the conjugacy clases of subgroups of G, with $r = r(G)$. As long as $\alpha_i > 1$, the factor $(N_G(H_i)/H_i) \wr S_{\alpha_i}$, in the decomposition of $\mathrm{ICA}(G; A)$, has a proper normal subgroup $(N_G(H_i)/H_i) \wr A_{\alpha_i}$ (where A_{α_i} is the alternating group of degree α_i). We know that $\alpha_i = 1$ if and only if $[G : H] = 2$ and $q = 2$ (Lemma 2). Hence, for $q \geq 3$, we have

$$\mathrm{Rank}(\mathrm{ICA}(G; A)) \geq \mathrm{Rank}\left(\frac{\prod_{i=1}^r ((N_G(H_i)/H_i) \wr S_{\alpha_i})}{\prod_{i=1}^r ((N_G(H_i)/H_i) \wr A_{\alpha_i})}\right) = \mathrm{Rank}\left(\prod_{i=1}^r \mathbb{Z}_2\right) = r.$$

Assume now that $q = 2$, and let $[H_1], \ldots, [H_{r_2}]$ be the conjugacy classes of subgroups of index two, with $r_2 = r_2(G)$. Now, $\mathrm{Rank}(\mathrm{ICA}(G; A))$ is at least

$$\mathrm{Rank}\left(\frac{\prod_{i=1}^r ((N_G(H_i)/H_i) \wr S_{\alpha_i})}{\prod_{i=1}^r ((N_G(H_i)/H_i) \wr A_{\alpha_i})}\right) = \mathrm{Rank}\left(\prod_{i=r_2+1}^r \mathbb{Z}_2\right) = r - r_2,$$

and the result follows. $\qquad\square$

The previous result could be refined for special classes of finite groups. In [5] this has been done for cyclic groups, and we do it next for dihedral groups.

Proposition 3. *Let $n \geq 1$ and A a finite set of size $q \geq 2$.*

$$\mathrm{Rank}(\mathrm{ICA}(D_{2n}; A)) \geq \begin{cases} \mathrm{d}_-(2n) + 2\mathrm{d}_+(2n) & \text{if } n \text{ is odd and } q \geq 3, \\ \mathrm{d}_-(2n) + 2\mathrm{d}_+(2n) - 1 & \text{if } n \text{ is odd and } q = 2, \\ \mathrm{d}_-(2n) + 2\mathrm{d}_+(2n) + 4\mathrm{d}_+(n) & \text{if } n \text{ is even and } q \geq 3, \\ \mathrm{d}_-(2n) + 2\mathrm{d}_+(2n) + 2\mathrm{d}_+(n) - 1 & \text{if } n \text{ is even and } q = 2, \end{cases}$$

Proof. We shall use the decomposition of $\mathrm{ICA}(D_{2n}; A)$ given in the proof of Theorem 2. For each $m_i \mid 2n$ even greater than 2, the corresponding α_i is greater than 1 by Lemma 2. The group $D_{m_i} \wr S_{\alpha_i}$ has a normal subgroup $N \cong (\mathbb{Z}_{m_i/2})^{\alpha_i}$ such that $(D_{m_i} \wr S_{\alpha_i})/N \cong \mathbb{Z}_2 \wr S_{\alpha_i}$. Now, $\mathbb{Z}_2 \wr S_{\alpha_i}$ has a normal subgroup

$$U = \left\{ ((a_1, \ldots, a_{\alpha_i}); \mathrm{id}) : \sum_{j=1}^{\alpha_i} a_j = 0 \mod (2) \right\}$$

such that $(\mathbb{Z}_2 \wr S_{\alpha_i})/U \cong \mathbb{Z}_2 \times S_{\alpha_i}$. Finally, a copy of the alternating group A_{α_i} is a normal subgroup of $\mathbb{Z}_2 \times S_{\alpha_i}$ with quotient group $\mathbb{Z}_2 \times \mathbb{Z}_2$. This implies that $D_{m_i} \wr S_{\alpha_i}$ has a normal subgroup with quotient group isomorphic to $\mathbb{Z}_2 \times \mathbb{Z}_2$.

Suppose that n is odd and $q \geq 3$. Then $\mathrm{ICA}(D_{2n}; A)$ has a normal subgroup with quotient group isomorphic to

$$\prod_{m_i \mid 2n \text{ odd}} \mathbb{Z}_2 \times \prod_{2 < m_i \mid 2n \text{ even}} (\mathbb{Z}_2)^2 \times (\mathbb{Z}_2)^2.$$

Thus, $\mathrm{d}_-(2n) + 2\mathrm{d}_+(2n) \leq \mathrm{Rank}(\mathrm{ICA}(D_{2n}; A))$. If $q = 2$, the last factor above becomes just $\mathbb{Z}_2$, as $\alpha_i = 1$ here, and the result follows.

Suppose that n is even and $q \geq 3$. Then $\mathrm{ICA}(D_{2n}; A)$ has a normal subgroup with quotient group isomorphic to

$$\prod_{m_i \mid 2n \text{ odd}} \mathbb{Z}_2 \times \prod_{2 < m_i \mid 2n \text{ even}} (\mathbb{Z}_2)^2 \times (\mathbb{Z}_2)^2 \times \prod_{m_i \mid n \text{ even}} (\mathbb{Z}_2)^4$$

Therefore, $\mathrm{d}_-(2n) + 2\mathrm{d}_+(2n) + 4\mathrm{d}_+(n) \leq \mathrm{Rank}(\mathrm{ICA}(D_{2n}; A))$. If $q = 2$, the last $\mathrm{d}_+(n) + 1$ factors become $\mathbb{Z}_2 \times \prod_{m_i \mid n \text{ even}} (\mathbb{Z}_2)^2$ and the result follows.   □

4.2   Infinite Groups

Now we turn our attention to the case when G is an infinite group. It was shown in [2] that $\mathrm{ICA}(\mathbb{Z}; A)$ (and so $\mathrm{CA}(\mathbb{Z}; A)$) is not finitely generated by studying its action on periodic configurations. In this section, using elementary techniques, we prove that the monoid $\mathrm{CA}(G; A)$ is not finitely generated when G is infinite abelian, free or infinite dihedral; this illustrates an application of the study of ranks of groups of cellular automata over finite groups.

Remark 1. Let G be a group that is not finitely generated. Suppose that $\mathrm{CA}(G; A)$ has a finite generating set $H = \{\tau_1, \ldots, \tau_k\}$. Let S_i be a memory set for each τ_i. Then $G \neq \langle \cup_{i=1}^{k} S_i \rangle$, so let $\tau \in \mathrm{CA}(G; A)$ be such that its minimal memory set is not contained in $\langle \cup_{i=1}^{k} S_i \rangle$. As a memory set for the composition $\tau_i \circ \tau_j$ is $S_i S_j = \{s_i s_j : s_i \in S_i, s_j \in S_j\}$, τ cannot be in the monoid generated by H, contradicting that H is a generating set for $\mathrm{CA}(G; A)$. This shows that $\mathrm{CA}(G; A)$ is not finitely generated whenever G is not finitely generated.

The next result, which holds for an arbitrary group G, will be our main tool.

Lemma 8. *Let G be a group and A a set. For every normal subgroup N of G,*

$$\mathrm{Rank}(\mathrm{CA}(G/N; A)) \leq \mathrm{Rank}(\mathrm{CA}(G; A)).$$

Proof. By [8, Proposition 1.6.2], there is a monoid epimorphism $\Phi : \mathrm{CA}(G; A) \rightarrow \mathrm{CA}(G/N; A)$. Hence, the image under Φ of a generating set for $\mathrm{CA}(G; A)$ of minimal size is a generating set for $\mathrm{CA}(G/N; A)$ (not necessarily of minimal size).   □

Theorem 5. *Let G be an infinite abelian group and A a finite set of size $q \geq 2$. Then, the monoid $\mathrm{CA}(G; A)$ is not finitely generated.*

Proof. If G is not finitely generated, then Remark 1 shows that $\mathrm{CA}(G; A)$ is not finitely generated, so assume that G is finitely generated. By the Fundamental Theorem of Finitely Generated Abelian Groups, G is isomorphic to

$$\mathbb{Z}^s \oplus \mathbb{Z}_{p_1} \oplus \mathbb{Z}_{p_2} \oplus \cdots \oplus \mathbb{Z}_{p_t},$$

where $s \geq 1$ (because G is infinite), and $p_1, \ldots, p_t$ are powers of primes. Then, for every $k \geq 1$, we may find a subgroup

$$N \cong \langle 2^k \rangle \oplus \mathbb{Z}^{s-1} \oplus \mathbb{Z}_{p_1} \oplus \mathbb{Z}_{p_2} \oplus \cdots \oplus \mathbb{Z}_{p_t}$$

such that $G/N \cong \mathbb{Z}_{2^k}$. By Lemma 8 and Proposition 2,

$$\text{Rank}(\text{CA}(G; A)) \geq \text{Rank}(\text{CA}(\mathbb{Z}_{2^k}; A)) \geq \text{Rank}(\text{ICA}(\mathbb{Z}_{2^k}; A)) \geq r(\mathbb{Z}_{2^k}) - 1 = k.$$

As the above holds for every $k \geq 1$, then $\text{CA}(G; A)$ is not finitely generated. $\square$

The *abelianization* of any group G is the quotient $G/[G,G]$, where $[G,G]$ is its *commutator subgroup*, i.e. the normal subgroup of G generated by all commutators $[g,h] := ghg^{-1}h^{-1}$, $g, h \in G$. The abelianization of G is in fact the largest abelian quotient of G.

Corollary 2. *Let G be a group with an infinite abelianization and A a finite set of size $q \geq 2$. Then, the monoid $\text{CA}(G; A)$ is not finitely generated.*

Proof. Let $G' = G/[G,G]$ be the abelianization of G. By Lemma 8, we have $\text{Rank}(\text{CA}(G'; A)) \leq \text{Rank}(\text{CA}(G; A))$. But $\text{CA}(G'; A)$ is not finitely generated by the previous theorem, so the result follows. $\square$

Corollary 3. *Let F_S be a free group on a set S and A a finite set of size $q \geq 2$. Then, the monoid $\text{CA}(F_S; A)$ is not finitely generated.*

Proof. As F_S has an infinite abelianization, which is the free abelian group on S, the result follows by the previous corollary. $\square$

The infinite dihedral group $D_\infty = \langle x, y \mid x^2 = y^2 = 1 \rangle$ has finite abelianization $\mathbb{Z}_2 \oplus \mathbb{Z}_2$. However, we can still show that $\text{CA}(D_\infty, A)$ is not finitely generated.

Proposition 4. *Let A be a finite set of size $q \geq 2$. Then, $\text{CA}(D_\infty; A)$ is not finitely generated.*

Proof. For every $n \geq 1$, define $H_n := \langle (xy)^n \rangle \leq D_\infty$, which is a normal subgroup of D_∞ with quotient group $D_\infty/H_n \cong D_{2n}$. By Proposition 2, As $r_2(D_n) = 1$, for every $n \geq 1$,

$$\text{Rank}(\text{CA}(D_\infty; A)) \geq \text{Rank}(\text{CA}(D_{2n}; A)) \geq r(D_{2n}) - 1,$$

We know that $r(D_{2n}) \geq \text{d}(2n)$, so, taking $n = 2^{k-1}$, $k \geq 1$, we see that $\text{Rank}(\text{CA}(D_\infty; A)) \geq d(2^k) - 1 \geq k$, for any $k \geq 1$. $\square$

Question 1. Is there an infinite group G such that $\text{CA}(G; A)$ is finitely generated?

The techniques of this section seem ineffective to answer this for infinite groups with few proper quotients, such as the infinite symmetric group.

Acknowledgments. The second author thanks the National Council of Science and Technology (CONACYT) of the Government of Mexico for the National Scholarship (No. 423151) which allowed him to do part of the research reported in this article.

References

1. Araújo, J., Schneider, C.: The rank of the endomorphism monoid of a uniform partition. Semigroup Forum **78**, 498–510 (2009)
2. Boyle, M., Lind, D., Rudolph, D.: The automorphism group of a shift of finite type. Trans. Am. Math. Soc. **306**(1), 71–114 (1988)
3. Cameron, P.J.: Permutation Groups. London Mathematical Society Student Texts 45. Cambridge University Press, Cambridge (1999)
4. Cameron, P.J., Solomon, R., Turull, A.: Chains of subgroups in symmetric groups. J. Algebra **127**(2), 340–352 (1989)
5. Castillo-Ramirez, A., Gadouleau, M.: Ranks of finite semigroups of one-dimensional cellular automata. Semigroup Forum **93**(2), 347–362 (2016)
6. Castillo-Ramirez, A., Gadouleau, M.: On finite monoids of cellular automata. In: Cook, M., Neary, T. (eds.) AUTOMATA 2016. LNCS, vol. 9664, pp. 90–104. Springer, Cham (2016). https://doi.org/10.1007/978-3-319-39300-1_8
7. Castillo-Ramirez, A., Gadouleau, M.: Cellular automata and finite groups. Nat. Comput. (2017). First Online
8. Ceccherini-Silberstein, T., Coornaert, M.: Cellular Automata and Groups. Springer Monographs in Mathematics. Springer, Berlin (2010). https://doi.org/10.1007/978-3-642-14034-1
9. Conrad, K.: Dihedral Groups II. http://www.math.uconn.edu/kconrad/blurbs/grouptheory/dihedral2.pdf
10. Gomes, G.M.S., Howie, J.M.: On the ranks of certain finite semigroups of transformations. Math. Proc. Camb. Phil. Soc. **101**, 395–403 (1987)
11. Gray, R.D.: The minimal number of generators of a finite semigroup. Semigroup Forum **89**, 135–154 (2014)
12. McIver, A., Neumann, P.: Enumerating finite groups. Q. J. Math. Oxford **38**(4), 473–488 (1987)

Enhancement of Automata with Jumping Modes

Szilárd Zsolt Fazekas[✉], Kaito Hoshi, and Akihiro Yamamura

Department of Mathematical Science and Electrical-Electronic-Computer Engineering, Akita University, 1-1 Tegata Gakuen-machi, Akita 010-8502, Japan
{szilard.fazekas,yamamura}@ie.akita-u.ac.jp, m8018308@s.akita-u.ac.jp

Abstract. Recently, new types of non-sequential machine models have been introduced and studied, such as jumping automata and one-way jumping automata. We study the abilities and limitations of automata with these two jumping modes of tape heads with respect to how they affect the class of accepted languages. We give several methods to determine whether a language is accepted by a machine with jumping mode. We also consider relationships among the classes of languages defined by the new machines and their classical counterparts.

Keywords: Jumping mode · Jumping finite automata · Pushdown automata · Pumping lemma · Context free language

1 Introduction

We study the ability of the jumping mode of tape heads to strengthen accepting power of automata. Automata have several characteristics; determinism of transition functions, ability of rewriting or erasing input words, memory devices like stacks and directions of tape heads to read inputs. Recently a mode of tape head movement has been introduced and examined with respect to how the class of languages accepted is affected ([1,3–6,8,9,11–13]). We study the abilities and limitations of the new mode of tape head move by comparing several machine models.

Jumping finite automata (JFA) were introduced as automata with a new mode of tape head in [12]. In the new mode, the tape head is allowed to jump - either left or right - over a part of the input word after reading a letter and continue processing from there. Once a letter in the input word is read, it cannot be reread again later. This implies that once a letter is read, it is erased. The tape head starts anywhere in the input word and it can move to either right or left side. It was shown that a language is accepted by jumping finite automata if and only if it is commutative and semilinear in [4,5].

One-way jumping (deterministic) finite automata (OWJFA), a variant of jumping finite automata, were introduced and analyzed in [4]. They have another mode of tape head; the head moves in one direction only and starts at the

© IFIP International Federation for Information Processing 2019
Published by Springer Nature Switzerland AG 2019
A. Castillo-Ramirez and P. P. B. de Oliveira (Eds.): AUTOMATA 2019, LNCS 11525, pp. 62–76, 2019.
https://doi.org/10.1007/978-3-030-20981-0_5

beginning of the input word. It moves from left to right (and jumps over parts of the input it cannot read) and when the tape head reaches the end of the input, it is returned to the beginning of the input and continues the computation until all the letters are read or the automaton is stuck in the sense that it can no longer read any letter of the remaining input. Several properties and characterization results were provided in [1]. The majority of decidability questions have been answered in [2], with the most notable exception being, when is the language accepted regular? We attempt to get closer to the answer by looking at the complements of OWJFA.

In this paper we consider deterministic or nondeterministic finite automata and pushdown automata in a uniform manner. However, we restrict our study to automata that do not rewrite the input letters and so we exclude linear bounded automata and Turing machines from consideration.

We recall notations of automata (see [7,15]). We denote $\Sigma_\epsilon = \Sigma \cup \{\epsilon\}$, where ϵ is the empty word and $P(Q)$ stands for the power set of Q. A *nondeterministic finite automaton* (denoted by NFA) M is a 5-tuple $(Q, \Sigma, \delta, q_0, F)$, where Q is set of states, Σ is finite set, $\delta : Q \times \Sigma \to P(Q)$ is a transition relation, q_0 is the initial state, F is set of accept state. In Sect. 3 we deviate from this definition by allowing the NFA to have multiple initial states, a change that is known not to affect the class of accepted languages in the classical case, but makes a difference in the alternative tape head modes. If δ is a mapping $Q \times \Sigma_\epsilon \to P(Q)$, then M is called ϵ−NFA. M is called *deterministic* (denoted by a DFA) if (1) it is an ε-free NFA and (2) for $\forall p \in Q$ and $\forall a \in \Sigma$, there is no more than one $q \in Q$ such that $\delta(p, a) = q$.

A *nondeterministic pushdown automaton* (denoted by NPDA) M is a 6-tuple $(Q, \Sigma, \Gamma, \delta, q_0, F)$, where Q is finite set of states, Σ is finite set, Γ is a finite stack alphabet, $\delta : Q \times \Sigma_\epsilon \times \Gamma_\epsilon \to P(Q \times \Gamma_\epsilon)$ is the transition function, q_0 is the initial state, F is set of accept states, and \$ is a bottom marker of stack. M is called *deterministic* (denoted by DPDA) if it satisfies (1) For $\forall q \in Q, \forall a \in \Sigma \cup \{\epsilon\}, \forall b \in \Gamma$ we have $|\delta(q, a, b)| \leq 1$ and (2) For $\forall q \in Q, \forall a \in \Sigma, \forall b \in \Gamma$, we have $\delta(q, a, b) = \emptyset$ if $\delta(q, \epsilon, b) \neq \emptyset$. The language accepted by DPDA is defined as the set of inputs on which the automaton ends up in a final state (not the empty stack acceptance condition) after reading the whole input.

2   Modes of Tape Head Move

First we define modes of movement of the tape head to capture characteristics of both JFA and (R)OWJFA. Transitions between configurations of automata are considered to be rewriting of strings on state and input alphabets. We study three ways of rewriting configurations of automata; the standard mode, the non-deterministic jumping mode and the one-way jumping mode. The first one is the traditional way to rewrite configurations of automata as defined in [7] and [15]. The second and third one are introduced by [12] and [4], respectively. The three modes can be applied to any automata with deterministic/nondeterministic transition functions, with/without stacks, and with/without rewriting and erasing a letter in an input. Our objective with this paper is to extend the examination of how the three modes relate to each other with respect to the accepting power of automata, in particular, nondeterministic finite and pushdown automata.

2.1 Tape Head Modes

Suppose M is a (deterministic or nondeterministic) finite automaton.

Standard Mode

A configuration of M is any string in $Q \times \Sigma^*$. A transition from configuration $q_1 aw$ to configuration $q_2 w$, written as $q_1 aw \rightarrow q_2 w$, is possible when $q_2 \in \delta(q_1, a)$. In the standard manner, we extend $\rightarrow$ to $\rightarrow^m$, where $m \geq 0$. Let $\rightarrow^+$ and $\rightarrow^*$ denote the transitive closure and the transitive-reflexive closure of $\rightarrow$, respectively.

We define a (deterministic or nondeterministic) finite automaton with *standard mode* to be a rewriting system $(M, \rightarrow)$ based on $\rightarrow^*$. The language accepted by $(M, \rightarrow)$ (denoted by $L(M, \rightarrow)$) is defined to be $\{w \mid w \in \Sigma^*, sw \rightarrow^* f, f \in F\}$.

Nondeterministic Jumping Mode

A configuration of M is any string in $\Sigma^* \times Q \times \Sigma^*$, representing the part of the input to the left from the reading head, the state and the input to the right from the head. The binary jumping relation, symbolically denoted by $\curvearrowright$, over $\Sigma^* \times Q \times \Sigma^*$, is defined as follows. Let x, z, x', z' be strings in Σ^* such that $xz = x'z'$ and $q \in \delta(p, y)$; then, M makes a jump from $xpyz$ to $x'qz'$, symbolically written as $xpyz \curvearrowright x'qz'$. In the standard manner, we extend $\curvearrowright$ to $\curvearrowright^m$, where $m \geq 0$. Let $\curvearrowright^+$ and $\curvearrowright^*$ denote the transitive closure of $\curvearrowright$ and the transitive-reflexive closure of $\curvearrowright$, respectively.

We define a (deterministic or nondeterministic) finite automaton with *nondeterministic jumping mode* to be a rewriting system $(M, \curvearrowright)$ based on $\curvearrowright^*$. (see Fig. 1). The language accepted by $(M, \curvearrowright)$ (denoted by $L(M, \curvearrowright)$) is defined to be $\{uv \mid u, v \in \Sigma^*, usv \curvearrowright^* f, f \in F\}$. This jumping mode was introduced as a new automaton model called a JFA in [12].

One-way Jumping Mode

The *right one-way jumping relation* (denoted by $\circlearrowright$) between configurations from $Q\Sigma^*$, was defined in [4] as follows. Suppose that x and y belong to Σ^*, a belongs to Σ, p and q are states in Q and $q \in \delta(p, a)$. Then the right one-way jumping automaton M makes a jump from the configuration $pxay$ to the configuration qyx, symbolically written as $pxay \circlearrowright qyx$ if x belongs to $\{\Sigma \setminus \Sigma_p\}^*$ where $\Sigma_p = \{b \in \Sigma \mid \exists q \in Q \ s.t. \ q \in \delta(p, b)\}$ (see Fig. 2). In the standard manner, we extend $\circlearrowright$ to $\circlearrowright^m$, where $m \geq 0$. Let $\circlearrowright^*$ denote the transitive-reflexive closure of $\circlearrowright$.

We define a (deterministic or nondeterministic) finite automaton with *one-way jumping mode of tape head* to be a rewriting system $(M, \circlearrowright)$ based on $\circlearrowright^*$. The language accepted by $(M, \circlearrowright)$ (denoted by $L(M, \circlearrowright)$) is defined to be $\{w \mid w \in \Sigma^*, sw \circlearrowright f, f \in F\}$.

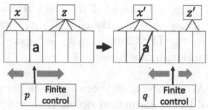

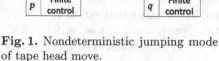

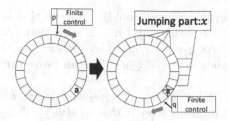

Fig. 1. Nondeterministic jumping mode of tape head move.

Fig. 2. One-way jumping mode of tape head move.

In a similar manner we define a (deterministic or nondeterministic) pushdown automaton with *standard mode, nondeterministic jumping mode* and *one-way jumping mode*, respectively, to be the rewriting systems $(M, \rightarrow)$, $(M, \curvearrowright)$ and $(M, \circlearrowleft)$, respectively, where M is a (deterministic or nondeterministic) pushdown automaton.

2.2 Language Classes

We consider deterministic and nondeterministic finite automata and deterministic and nondeterministic pushdown automata, denoted by DFA, NFA, DPDA, NPDA, respectively. Then we classify these automata with three modes of tape head from the standpoint of languages accepted. We denote the language classes accepted by DFA, NFA, DPDA, NPDA with three modes $\rightarrow, \curvearrowright, \circlearrowleft$ by $(\rightarrow, \curvearrowright, \circlearrowleft)$-**DFA**, $(\rightarrow, \curvearrowright, \circlearrowleft)$-**NFA**, $(\rightarrow, \curvearrowright, \circlearrowleft)$-**DPDA**, $(\rightarrow, \curvearrowright, \circlearrowleft)$-**NPDA**, respectively, in this paper. For example, $\rightarrow$**DFA** coincides with $\rightarrow$**NFA** and they comprise the class of regular languages, and $\rightarrow$**NPDA** is the class of context-free languages.

2.3 Differences of Modes of Tape Head Move

The next proposition shows the basic relationship between the processing of inputs by the same machine M in different tape head modes. Versions of the statement regarding the acceptance of inputs by M in the different modes have been shown in [13, Ch.17] and [4], for $\curvearrowright$ and $\circlearrowleft$, respectively.

Proposition 1. *Let M be an (deterministic or non-deterministic) automaton. Suppose w_1, w_2, w_3 are words over Σ.*
(1) If $q_1 w_1 \rightarrow^* q_2 w_2$ then $q_1 w_1 \curvearrowright^* q_2 w_2$.
(2) If $q_1 w_1 \rightarrow^* q_2 w_2$ then $q_1 w_1 \circlearrowleft^* q_2 w_2$.
(3) If $q_1 w_1 \curvearrowright^* q_2 w_2$ then there exists a permutation ϕ such that $q_1 \phi(w_1) \rightarrow^* q_2 w_2$.
(4) If $q_1 w_1 \circlearrowleft^* q_2 w_2$ then there exists a permutation ϕ such that $q_1 \phi(w_1) \rightarrow^* q_2 w_2$.

Proof. (1) Let $q_1 w_1 \rightarrow^n q_2 w_2$ and let $m = |w_1|$. We prove that $q_1 w_1 \curvearrowright^n q_2 w_2$ holds, by induction on n. If $n = 0$, then $q_1 = q_2$ and $w_1 = w_2$. Thus,

$q_1w_1 \curvearrowright^0 q_2w_2$ holds. Suppose that $n = k \le m - 1$ holds. If $n = k+1$, then there exist q and $a \in \Sigma$ such that $q_1w_1 \to^k qaw_2 \to q_2w_2$. By $q_2 \in \delta(q,a)$, we have $q_1w_1 \curvearrowright^k qaw_2 \curvearrowright q_2w_2 = q_1w_1 \curvearrowright^{k+1} q_2w_2$. Therefore, if $q_1w_1 \to^* q_2w_2$, then $q_1w_1 \curvearrowright^* q_2w_2$. We can prove (2) in a similar manner.

(3) Let $q_1w_1 \curvearrowright^n q_2w_2$ and let $m = |w_1|$. We prove that there exists a permutation ϕ such that $q_1\phi(w_1) \to^n q_2w_2$ holds, by induction on n. If $n = 0$, then there exists a permutation ϕ such that $\phi(w_1) = w_1$. Thus, $q_1\phi(w_1) \to^0 q_1w_1 = q_2w_2$ holds. Suppose that $n = k \le m - 1$ holds. i.e. there exists $\sigma = \begin{pmatrix} 1 & 2 & 3 & \cdots & k-1 & k \\ \sigma(1) & \sigma(2) & \sigma(3) & \cdots & \sigma(k-1) & \sigma(k) \end{pmatrix}$ such that $q_1\sigma(w_1) \to^k q_2w_2$. If $n = k+1$, then there exist q and $a \in \Sigma$ such that $q_1w_1 \curvearrowright^k quav \curvearrowright q_2w_2$ where $uv = w_2$. From before, $q_1\sigma(w_1) \to^k quav \curvearrowright q_2w_2$ holds and let the permutation σ' be such that $\sigma'(xyaz) = xayz$ for $a \in \Sigma$ and $xyz \in \Sigma^*$ with $|x| = k$. Choosing $\phi = \sigma' \circ \sigma$ gives $q_1\phi(w_1) \to^k qauv \to q_2w_2$. Thus, $q_1\phi(w_1) \to^{k+1} q_2w_2$ holds.
We can prove (4) in a similar manner.   □

Remark 1. It is easy to see that if $q_1w_1w_2 \to^* q_3$ and $q_1w_1 \to^* q_2$ then $q_2w_2 \to^* q_3$. However, $q_1w_1w_2 \curvearrowright^* q_3$ and $q_1w_1 \curvearrowright^* q_2$ does not imply $q_2w_2 \curvearrowright^* q_3$. Similarly, $q_1w_1w_2 \circlearrowleft^* q_3$ and $q_1w_1 \circlearrowleft^* q_2$ does not imply $q_2w_2 \circlearrowleft^* q_3$.

In [1](equation (1)) the authors summarize the relationship between the languages accepted by DFA M in the various tape head modes, as:

$$L(M, \to) \subseteq L(M, \circlearrowleft) \subseteq L(M, \curvearrowright) = Perm(L(M, \to)).$$

Regarding the $\to$ and $\curvearrowright$ modes, it makes no difference in the proofs whether the machine is deterministic. This means that we have the following relationship between acceptance in $\to$ and $\curvearrowright$ modes.

Proposition 2. *Let M be any automaton. A word w is accepted by $(M, \curvearrowright)$ if and only if there exists a permutation ϕ_w depending on w such that $\phi_w(w)$ is accepted by $(M, \to)$.*

We note that the permutation ϕ_w depends on the word w, however, it is not necessarily unique. Therefore, if a language L is accepted by a JFA $(M, \curvearrowright)$, we have $L = Perm(L)$.

Corollary 1. *A word w is accepted by a rewriting system $(M, \curvearrowright)$ if and only if there exists a permutation ϕ such that $\phi(w)$ is accepted by a rewriting system $(M, \to)$.*

The following corollary has been stated for JFA ($\curvearrowright$NFA) in [12] and for deterministic ROWJFA ($\circlearrowleft$DFA) in [4]. They also follow, just as the nondeterministic case, from Proposition 1. A similar argument can easily be made for DPDA and NPDA.

Corollary 2. *Let M be a deterministic or nondeterministic finite automaton or pushdown automaton. Any word w, which is accepted by $(M, \to)$, is also accepted by $(M, \curvearrowright)$ and $(M, \circlearrowleft)$.*

This means, generalizing the inclusions from before, that for any DFA, NFA, DPDA or NPDA M:

$$L(M, \rightarrow) \subseteq L(M, \circlearrowright) \subseteq L(M, \curvearrowright) = Perm(L(M, \rightarrow)). \qquad (1)$$

Parikh's well-known paper [14] shows that all context-free languages have semilinear Parikh-image. Furthermore, all semilinear sets have a regular preimage, and this, combined with Eq. (1) tell us that the class of permutation closures of context-free languages is the same as the class of permutation closures of regular languages, accepted by DFA/NFA in $\curvearrowright$ mode ([5]). From here, by Corollary 1, we can deduce that in $\curvearrowright$ tape head mode, adding a stack to a finite automaton does not change the class of accepted languages, i.e.,

$$\curvearrowright \textbf{DFA} = \curvearrowright \textbf{DPDA} = \curvearrowright \textbf{NPDA}.$$

Equation (1) also allows us to extend all the well-known pumping lemmas trivially to the language classes defined through the new execution modes.

Remark 2. Let M be an NFA, DFA, DPDA or NPDA, and $N_M > 0$ a constant which depends on M. If a pumping lemma holds for all words $w \in L(M, \rightarrow)$ with $|w| \geq N_M$, then for all words $w \in L(M, \curvearrowright) \cup L(M, \circlearrowright)$ with $|w| \geq N_M$, there exists a permutation ϕ such that the lemma holds for $\phi(w)$.

To close this section, we briefly discuss the special cases of unary and binary alphabets for the new tape head modes.

Theorem 1. *Let M be a deterministic or nondeterministic finite automaton or pushdown automaton with $|\Sigma| = 1$, then $L(M, \curvearrowright) = L(M, \circlearrowright) = L(M, \rightarrow)$.*

Proof. Straightforward, as the languages above are commutative and semilinear. ☐

It follows that there is no $\circlearrowright$NPDA that accepts $\{a^p \mid \text{p is a prime number}\}$, providing separation of $\circlearrowright\textbf{NPDA}$ from the class of context-sensitive languages. Binary alphabets are also special cases for the $\curvearrowright$ mode. As shown in [10], over binary alphabets all commutative semilinear languages are context-free, so we have $\curvearrowright\textbf{NFA} \subset \rightarrow\textbf{NPDA}$. This is in contrast with larger alphabets, where $\curvearrowright\textbf{NFA}$ and $(\rightarrow)$-$\textbf{NPDA}$ are incomparable.

3   One-Way Jumping Nondeterministic Finite Automata ($\circlearrowright$NFA)

In previous papers [1, 2, 4] the class $\circlearrowright\textbf{DFA}$ of languages accepted by one-way jumping deterministic finite automata has been investigated. In [1] it was shown that permutation closed languages in $\circlearrowright\textbf{DFA}$ are characterized by having finitely many positive Myhill-Nerode equivalence classes. This nice characterization gave the corollary that a permutation closed language in $\circlearrowright\textbf{DFA}$ is regular if and only if its complement is in $\circlearrowleft$**DFA**. However, not much more has been known about the complements of $\circlearrowleft$**DFA** languages. To tackle this, we look at the class $\circlearrowleft$**NFA** of languages accepted by one-way jumping nondeterministic finite automata ($\circlearrowleft$NFA), that is, machines $(M, \circlearrowleft)$, where M is an NFA without ϵ-moves, but with possibly multiple initial states. Then, we show that $\circlearrowleft$**NFA** strictly includes $\curvearrowright$**NFA**, the class of permutation closed semilinear languages and, as a corollary, the complements of permutation closed $\circlearrowleft$**DFA** languages. Finally, we go further by showing that $\circlearrowleft$**NFA** contains the complements of all $\circlearrowleft$**DFA** languages, but $\circlearrowleft$**NFA** itself is not closed under complementation.

Definition 1. *($\circlearrowleft$NFA) A (right) one-way jumping nondeterministic finite automaton is an ϵ-free NFA with multiple initial states in $\circlearrowleft$ execution mode.*

Example 1. The language $K = \{w \in \{a,b\}^* : |w|_b = 0 \text{ or } |w|_a = |w|_b\}$ is accepted by $\circlearrowleft$NFA $M = (\{q_0, q_1, q_2, q_3, q_4, q_5\}, \{a, b\}, \delta, \{q_0\}, \{q_1, q_3\})$, where $\delta(q_0, a) = \{q_1, q_3\}, \delta(q_1, a) = \{q_2\}, \delta(q_2, b) = \{q_1\}, \delta(q_3, a) = \{q_3\}$.

In [1] it is shown that $K \notin \circlearrowleft$**DFA**; this establishes $\circlearrowleft$**DFA**$\subsetneq\circlearrowleft$**NFA**. To move on to the inclusions in $\circlearrowleft$**NFA** of the classes mentioned above, first we need to state that, as expected, $\circlearrowleft$**NFA** is closed under union, by a construction similar to the case of classical NFA.

Proposition 3. *The class $\circlearrowleft$**NFA** is closed under union.*

Proof. Let $M_1 = (Q_1, \Sigma_1, \delta_1, S_1, F_1)$ and $M_2 = (Q_2, \Sigma_2, \delta_2, S_2, F_2)$ be two NFA such that $Q_1 \cap Q_2 = \emptyset$ (if this does not hold, we can simply rename the states). It is straightforward to see that $M_3 = (Q_1 \cup Q_2, \Sigma_1 \cup \Sigma_2, \delta_1 \cup \delta_2, S_1 \cup S_2, F_1 \cup F_2)$ accepts the union of the languages accepted by M_1 and M_2 in $\circlearrowleft$ execution mode, that is, $L(M_3, \circlearrowleft) = L(M_1, \circlearrowleft) \cup L(M_2, \circlearrowleft)$. $\qquad\square$

Theorem 2. *For any semilinear set $S \in \mathbb{N}^{|\Sigma|}$ there exists an NFA M, such that $\Psi_\Sigma^{-1}(S) = L(M, \circlearrowleft)$, where Ψ_Σ is the Parikh mapping for alphabet Σ.*

Proof. S is semilinear, so we can write it as the finite union of linear sets. We give the construction for a linear set. The statement then follows from Proposition 3, by constructing disjoint NFA for all linear sets and taking their union to be M. Let $\Sigma = \{a_1, \ldots, a_k\}$ and $Lin \subseteq \mathbb{N}^k$ be a linear set of vectors over non-negative integers, i.e., there exist $v_0, \ldots, v_n \in \mathbb{N}^k$ such that

$$Lin = \{v_0 + c_1 v_1 + \cdots + c_n v_n \mid c_1, \ldots, c_n \in \mathbb{N}\}.$$

First we define the set of "starting vectors" as $Lin_0 = \{v_0, v_0 + v_1, \ldots, v_0 + v_n\}$, the minimal subset of Lin such that all letters which occur in Lin, also occur in Lin_0. The set $W(Lin_0)$ of representative words for a set of vectors Lin_0 will be a subset of the preimage of Lin_0 under the Parikh mapping. For each $(m_1, \ldots, m_k) \in Lin_0$ we take k words such that all possible starting letters are represented (there may be less than k if some $m_i = 0$). Several choices for $W : \mathbb{N}^{|\Sigma|} \to \Sigma^*$ are possible. We set W such that

$$W(Lin_0) = \bigcup_{(m_1,\ldots,m_k)\in Lin_0} \{a_1^{m_1}a_2^{m_2}\cdots a_k^{m_k}, a_2^{m_2}\cdots a_k^{m_k}a_1^{m_1}, \ldots, a_k^{m_k}a_1^{m_1}\cdots a_{k-1}^{m_{k-1}}\}$$

Now we construct M as follows. Let the initial state of M be s and for each $w_i \in W(Lin_0)$ add a new path from s to a new final state w_i labeled by letters of w_i (see Fig. 3). To each of these new final states w_i we add all possible loops labeled by words in $W(\{v_1, \ldots, v_n\})$ (see Fig. 4). Here we removed v_0, because it is only added once to each vector in Lin, according to the definition. It is clear from the construction that for each vector $v \in Lin$ there is at least a path in M labeled by a word in the preimage of v. Conversely, the labels of each path form a word whose Parikh mapping is some $v \in Lin$. Even though not every preimage of v forms a path in M, the $\circlearrowright$ mode of execution allows M to read all the letters.   □

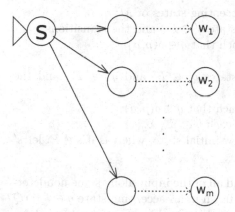

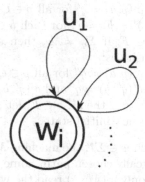

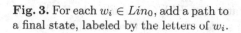

Fig. 3. For each $w_i \in Lin_0$, add a path to a final state, labeled by the letters of w_i.

Fig. 4. For each $u_j \in W(\{v_1, \ldots, v_n\})$ add a loop to each w_i, labeled by the letters of u_j.

At first sight it may seem that adding multiple paths for each vector in a linear set to the NFA in the previous proof is an overkill. While in some cases this could be avoided, as the next example shows, it is necessary in general.

Example 2. Consider the semilinear set, which is the union of linear sets $Lin_1 = \{(1,0,0,0) + c_1 \cdot (0,2,2,0) + c_2 \cdot (0,0,2,2) + c_3 \cdot (0,2,0,2) \mid c_1, c_2, c_3 \in \mathbb{N}\}$ and $Lin_2 = \{(0,1,0,0) + c_1 \cdot (2,0,0,0) + c_2 \cdot (2,2,0,0) \mid c_1, c_2 \in \mathbb{N}\}$ over alphabet $\{a, b, c, d\}$. If in the first phase of the construction only v_0 was to be added as a path from s for each linear set, then for input $bbcca$, the machine would have to choose the branch of Lin_2, because the first letter to the right for which there is a transition from s is b. This would mean $bbcca$ is rejected, because $(1, 2, 2, 0) \notin Lin_2$, even though $(1, 2, 2, 0) \in Lin_1$ and thus $bbcca \in \Psi^{-1}(Lin_1 \cup Lin_2)$. If in the second phase of the construction only one path per vector v_i were to be added to the states w_i, then one of the inputs $abbcc$, $accdd$ or $addbb$ would be rejected even though they are in the language.

Corollary 3. $\curvearrowright$NFA$\subset\circlearrowleft$NFA.

Proof. As it was shown in [4], $\curvearrowright$**NFA** is the class of permutation closed semilinear languages, that is, the class of languages which are preimages of semilinear sets under the Parikh mapping. $\square$

Theorem 3. *For any DFA $M = (Q, \Sigma, \delta, s, F)$, we can construct an NFA M' such that $\Sigma^* \setminus L(M, \circlearrowleft) = L(M', \circlearrowleft)$.*

Proof. As we mentioned before, Σ_q denotes the set of letters for which there is an outgoing transition from $q \in Q$. We construct the NFA $M' = (Q', \Sigma, \delta', \{s, s'\}, F')$ by the following steps:

1. switch the accepting states and non-accepting states of M;
2. $\forall q \in Q \setminus F$: if $\Sigma_q \neq \Sigma$, then add new state $q' \in F'$, and the transitions:
    - $\delta'(p, a) = q'$ for all $p \in Q, a \in \Sigma$ such that $q \in \delta(p, a)$
    - $\delta'(q', b) = q'$, for each $b \in \Sigma \setminus \Sigma_q$;
3. $\forall q \in F$: if $\Sigma_q \neq \Sigma$, then add new states $q' \in F'$ and $q'' \notin F'$, and the transitions:
    - $\delta'(p, a) = q''$ for all $p \in Q, a \in \Sigma$ such that $q \in \delta(p, a)$
    - $\delta'(q'', b) = q'$ and $\delta'(q', b) = q'$, for each $b \in \Sigma \setminus \Sigma_q$;
4. if $\Sigma_s \neq \Sigma$ then if $s \in F$, let s'' be a new initial state, whereas if $s \notin F$, let s' be a new initial state.

For each $w \in \Sigma^*$, the machine M' will read all of the input along some nondeterministically chosen path. Some path will finish in an accepting state $q \in F' \cap Q$ if and only if $(M, \circlearrowleft)$ read the whole input and stopped in a non-accepting state. Some path will finish in an accepting state $q' \in F' \setminus Q$ if and only if $(M, \circlearrowleft)$ could not read the input and got stuck in state q, because the remaining letters were all elements of Σ_q. In more detail, for each input w one of the following happens:

- $w \in L(M, \circlearrowleft)$: this means M reads all the input and finishes in a state from $q \in F$. The only branches of M' which read the whole input finish in q or q'' (if it exists), but since $q, q'' \notin F'$, the input is rejected by M';
- $w \notin L(M, \circlearrowleft)$ and M reads the whole input finishing in some state $q \notin F$. In this case, the same path in M' reads the whole input and finishes in $q \in F'$, so M' accepts the input;
- $w \notin L(M, \circlearrowleft)$ and M cannot read the whole input, so M gets stuck in a state q, with the remaining input being in $(\Sigma \setminus \Sigma_q)^*$. Depending on whether q is final or not, we distinguish two subcases:

- $q \in F$: from the states preceding q in the path, the branch that goes to q gets stuck. The branch that goes to q'' reads the remaining input after transitioning to $q' \in F'$, so M' accepts;
- $q \notin F$: from the states preceding q in the path, the branch that goes to q gets stuck. The branch that goes to $q' \in F'$ reads the remaining input, so M' accepts. □

Example 3. To illustrate how the construction works (Fig. 5), consider the non-regular $\circlearrowleft$**DFA** language $L_{ab} = \{w \in \{a,b\}^* \mid |w|_a = |w|_b\}$, which can be accepted by the $\circlearrowleft$DFA $M = (\{s,q\}, \{a,b\}, \delta, s, \{s\})$, where $\delta(s,a) = q$ and $\delta(q,b) = s$.

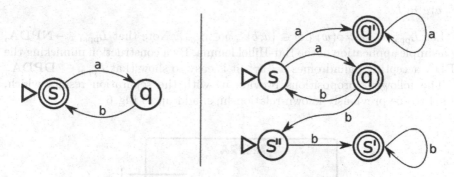

Fig. 5. Left: $\circlearrowleft$DFA for $L_{ab} = \{w \in \{a,b\}^* \mid |w|_a = |w|_b\}$. Right: $\circlearrowleft$NFA for $\Sigma^* \setminus L_{ab}$.

4 One-Way Jumping Pushdown Automaton: ($\circlearrowleft$)-NPDA, ($\circlearrowleft$)-DPDA

In this section we initiate the study of how the $\circlearrowleft$ tape head mode affects the computational power of PDA. In particular, we exhibit certain languages which provide separation between the classes of languages accepted by DPDA and NPDA in the $\circlearrowleft$ mode and their classical counterparts, as well as, finite automata in $\circlearrowleft$ mode. We also present languages which show that several common closure properties do not apply for $\circlearrowleft$**NPDA**. The proofs for the results that follow are based on extended versions of pumping lemmas for deterministic context-free languages ([16]) and context-free languages (Bar-Hillel).

First, we state two extended pumping lemmas for $\circlearrowleft$ mode. The proofs of these lemmas are trivial based on Remark 2.

Corollary 4 (Bar-Hillel lemma for $\circlearrowleft$-NPDA). *For any language L accepted by a $\circlearrowleft NPDA$ there exists a constant n, such that for every string $w \in L$ with $|w| > n$, there exists a permutation w_σ, which can be written as $w_\sigma = uvxyz$, satisfying (1) $|vy| \geq 1$, (2) $|vxy| \leq n$ and (3) $uv^i xy^i z \in L$ for every $i \geq 0$.*

Corollary 5 (Pumping Lemma for $\circlearrowright$DPDA, original version in [16]).
Suppose L is accepted by a $\circlearrowright$DPDA M. Then there exists a constant n for L such that for any pair of words $w, w' \in L$ if

(1) $s = xy$ and $s' = xz$, $|x| > n$, and
(2) (first symbol of y) = (first symbol of z),

where s and s' are permutations of w and w', such that $s, s' \in L(M, \to)$, then either (3) or (4) holds:

(3) there is a factorization $x = x_1 x_2 x_3 x_4 x_5$, $|x_2 x_4| \geq 1$ and $|x_2 x_3 x_4| \leq n$, such that for all $i \geq 0$, $x_1 x_2{}^i x_3 x_4{}^i x_5 y$ and $x_1 x_2{}^i x_3 x_4{}^i x_5 z$ are in L;

(4) there exist factorizations $x = x_1 x_2 x_3$, $y = y_1 y_2 y_3$ and $z = z_1 z_2 z_3$, $|x_2| \geq 1$ and $|x_2 x_3| \leq n$, such that for all $i \geq 0$, $x_1 x_2{}^i x_3 y_1 y_2{}^i y_3$ and $x_1 x_2{}^i x_3 z_1 z_2{}^i z_3$ are in L.

Let $L_{\mathrm{ppal}} = \{w \# \phi(w) \mid w \in \{a, b\}^*, \phi \in S_{|w|}\}$. Note that $L_{\mathrm{ppal}} \notin {\to}\mathbf{NPDA}$, by a simple application of the Bar-Hillel lemma. By a construction mimicking the DPDA accepting palindromes $w \# w^R$, it is easy to show that $L_{\mathrm{ppal}} \in \circlearrowright\mathbf{DPDA}$.

The following propositions provide us with the separation results which, added to the previously known relationships, add up to Fig. 6.

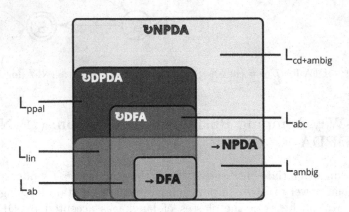

Fig. 6. Relationship between REG=${\to}$DFA=${\to}$NFA, CF=${\to}$NPDA, and one-way jumping classes.

Proposition 4. $L_{\mathrm{ppal}} \notin \circlearrowright\mathbf{DFA}$.

Proof. Suppose that $M = (Q, \Sigma, \delta, q_0, F)$ is a DFA, such that $L(M, \circlearrowright) = L_{\mathrm{ppal}}$. Consider the word $w = a^p \# a^p \in L_{\mathrm{ppal}}$, with $p = |Q| + 1$. By Proposition 1 there exists a permutation P such that $P(w) \in L(M, \to) \subseteq L_{\mathrm{ppal}}$ and the pumping lemma for regular languages says that it can be written as $P(w) = xyz$, where $y \neq \epsilon, |xy| \leq |Q|$ and $xy^i z \in L_{\mathrm{ppal}}, \forall i \geq 0$. Since $L_{\mathrm{ppal}} \cap (a + \#)^*$ has only one word of each length, we have $P(w) = a^p \# a^p$. From $|xy| \leq |Q|$, we get a contradiction when $i = 2$ as $xy^i z \notin L_{\mathrm{ppal}}$. $\qquad\square$

Proposition 5. $L_{ambig} = \{a^i b^i | i \geq 0\} \cup \{a^i b^{2i} | i \geq 0\} \notin \circlearrowleft\mathbf{DPDA}$.

Proof. Assume there exists DPDA $M = (Q, \Sigma, \Gamma, \delta, q_0, F)$ such that $L(M, \circlearrowleft) = L_{ambig}$ and let C be the constant for L_{ambig} in Corollary 5.

Choose $w = a^n b^n$ and $w' = a^n b^{2n}$ for some integer $n > C$, then there exist permutations σ and σ' such that $w_\sigma, w'_{\sigma'} \in L(M, \rightarrow)$. By $w_\sigma, w'_{\sigma'} \in L(M, \rightarrow) \subseteq L(M, \circlearrowleft) = L_{ambig}$, we get that $w_\sigma = a^n b^n$ and $w'_{\sigma'} = a^n b^{2n}$. Let $x = a^n b^{n-1}$, $y = b$, and $z = b^{2n-1}$. The choice of $w_\sigma = xy$ and $w'_{\sigma'} = xz$ satisfies (1) and (2) of Corollary 5. According to Corollary 5, either (3) or (4) should hold.

Let us consider (3) first. The only possible factorization $x = x_1 x_2 x_3 x_4 x_5$ such that $|x_2 x_4| > 0$ and for all i, $x_1 x_2{}^i x_3 x_4{}^i x_5 y \in L_{ambig}$ must satisfy the condition $x_2 = a^k$ and $x_4 = b^k$ for some $k > 0$. But then $x_1 x_2{}^0 x_3 x_4{}^0 x_5 z = x_1 x_3 x_5 z = a^{n-k} b^{2n-k} \notin L_{ambig}$. Therefore (3) does not hold.

Now, we consider (4). Any factorization $x = x_1 x_2 x_3$ such that $|x_2| > 0$ and $|x_2 x_3| \leq C < n$ will result in $x_2 \in b^+$ and $y_2 \in b^*$, so $x_1 x_3 y_1 y_3 = a^n b^{n-|x_2|-|y_2|} \notin L_{ambig}$. So (4) does not hold either.

This contradicts the $\circlearrowleft$-DPDA pumping lemma, so $L_{ambig} \notin \circlearrowleft\mathbf{DPDA}$. $\square$

The language L_{ambig} is a classic example of nondeterministic context-free language. At the same time, as mentioned in [4],

$$L_{abc} = \{w \in \{a, b, c\}^* \mid |w|_a = |w|_b = |w|_c\} \in \circlearrowleft\mathbf{DFA}\backslash\rightarrow\mathbf{NPDA}.$$

It is also straightforward that

$$L_{lin} = \{a^n b^n \mid n > 0\} \in \circlearrowleft\mathbf{DPDA}\cap\rightarrow\mathbf{NPDA},$$

and $L_{lin} \notin \circlearrowleft\mathbf{DFA}$, by [4, Cor. 12]. Since $\rightarrow\mathbf{NPDA}$ is trivially included in $\circlearrowleft\mathbf{NPDA}$, by the same arguments as above, the union $L_{cd} \cup L_{ambig} = L_{cd+ambig}$, where $L_{cd} = \{w \in \{c, d\}^* \mid |w|_c = |w|_d\}$, is in $\circlearrowleft\mathbf{NPDA}$. At the same time, it is easy to see that $L_{cd+ambig} \notin (\rightarrow\mathbf{NPDA}\cup\circlearrowleft\mathbf{DPDA})$. This completes the separation examples for Fig. 6. Finally, let us say a few words about closure properties of $\circlearrowleft\mathbf{NPDA}$.

Proposition 6. $L_1 = \{a^n b^n c^n | n \geq 0\} \notin \circlearrowleft\mathbf{NPDA}$.

Proof. Suppose that $M = (Q, \{a, b, c\}, \Gamma, \delta, q_0, F)$ and $L(M, \circlearrowleft) = L_3$. Consider the word $w = a^n b^n c^n \in L_1$. By Corollary 4 there exists a permutation σ such that $w_\sigma \in L_1$ can be written as $w_\sigma = uvxyz$, where $|vy| \geq 1$, $|vxy| \leq n$, $uv^i xy^i z \in L_1$, for all $i \geq 0$, where n is the contant from the Bar-Hillel lemma for $\circlearrowleft\mathbf{NPDA}$. Depending on the decomposition $uvxyz$, we have two cases

1. If vxy is generated by one symbol, then $uv^2 xy^2 z$ does not include the same number of a, b, c. This contradicts $uv^i xy^i z \in L_1$, for $i = 2$.
2. If vxy contains two kinds of symbols, then $uv^2 xy^2 z \notin L_1$, because the number of copies of the third letter (the one not in vxy) does not match the other two. This contradicts $uv^i xy^i z \in L_1$, when $i = 2$.

Therefore, no $\circlearrowleft\mathbf{NPDA}$ accepts L_1. $\square$

Since both $L_{abc} = \{w \mid |w|_a = |w|_b = |w|_c\} \in \circlearrowleft\textbf{NPDA}$ and $a^*b^*c^* \in \circlearrowleft\textbf{NPDA}$, from Proposition 6 we get that $\circlearrowleft\textbf{NPDA}$ is not closed under intersection. Together with the fact that it is closed under union, we get that it is not closed under complementation, either.

Theorem 4. $L_2 = \{wa \mid |w|_a = |w|_b = |w|_c\} \notin \circlearrowleft\textbf{NPDA}$.

Proof. Suppose that there exist a $\circlearrowleft\textbf{NPDA}$ $M = (Q, \{a, b, c\}, \Gamma, \delta, q_0, F)$, which accepts L_2. Consider the words $w_1 = a^m b^m c^m a \in L_2$ and $w_2 = a^{m+1} b^m c^m \notin L_2$, where m is the constant from Corollary 4. By $w_1 = a^m b^m c^m a \in L_2$, there exists permutation P such that $P(w_1) \in L(M, \rightarrow)$. Let $P(w_1) = x_1 x_2 ... x_{3m+1}$, then $x_{3m+1} = a$ by $P(w_1) \in L_2$. We will use the fact that w_2 is the cyclic shift of w_1. Let us look at the computation performed by the automaton on reading w_1. Before reading b or c, the automaton will read a^k for some $k \geq 0$, that is:

$$(q_0, a^m b^m c^m a, \$) \circlearrowleft (q_1, a^{m-1} b^m c^m a, z_1) \circlearrowleft .... \circlearrowleft (q_k, a^{m-k} b^m c^m a, z_k)$$

for some $\{q_1, q_2 ... q_k\} \subset Q$ and $\{z_1, z_2 .... z_k\} \subset \Gamma^*$ ($\$$ is the bottom marker for the pushdown). Since $w \in L$, there is an accepting computation from the last configuration $(q_k, a^{m-k} b^m c^m a, z_k)$ in that sequence. Depending on k, we have two cases.

(CASE 1) if $k < m$: we get that there is no transition defined from state q_k on reading a, therefore $a^{m-k}, a^{m-k+1} \in (\Sigma - \Sigma_{(q_k, z_k)})^*$, where $\Sigma_{(q_k, z_k)} = \{d \in \Sigma \mid \delta(q_k, d, z_k) \neq \emptyset\}$. We get that

$$(q_k, uv, z_k) = (q_k, a^{m-k+1} b^m c^m, z_k) \circlearrowleft^* (q, \epsilon, z)$$

if and only if

$$(q_k, vu, z_k) = (q_k, b^m c^m a^{m-K+1}, z_k) \circlearrowleft^* (q, \epsilon, z).$$

This means $(q_0, a^{m+1} b^m c^m, \$) \circlearrowleft^* (q_k, a^{m-k+1} b^m c^m, z_k) \circlearrowleft^* (q, \epsilon, z)$, which contradicts the initial assumption $a^{m+1} b^m c^m \notin L_2$.

(CASE 2) if $k = m$: This case says that $(q_0, a^m b^m c^m a, \$) \circlearrowleft^k (q_m, b^m c^m a, z_m)$.

Let $\circlearrowleft\textbf{NPDA}$ M then read b. Let us look at the computation performed by the automaton on reading $b^m c^m a$. Before reading c, the automaton will read b^l for some $l \geq 0$, that is:

$$(q_m, b^m c^m a, z_m) \circlearrowleft (q_{m+1}, b^{m-1} c^m a, z_{m+1}) \circlearrowleft .... \circlearrowleft (q_{m+l}, b^{m-l} c^m a, z_{m+l})$$

for some $\{q_{m+1}, q_{m+2} ... q_{m+l}\} \subset Q$ and $\{z_{m+1}, z_{m+2} .... z_{m+k}\} \subset \Gamma^*$. There is an accepting computation from the last configuration $(q_{m+l}, b^{m-l} c^m a, z_{m+l})$ in that sequence. Depending on l, we have two cases.

(CASE 2-1) if $l < m$: This case is the same as (CASE 1).

(CASE 2-2) if $l = m$: This case is the same as (CASE 2). $\circlearrowleft$NPDA M must read c. Let us look at the computation performed by the automaton on reading $c^m a$. Before reading the last letter, a, the automaton will read c^n for some $n \geq 0$:

$$(q_{2m}, c^m a, z_{2m}) \circlearrowleft^n (q_{2m+n}, c^{m-n} a, z_{2m+n})$$

for some $q_{2m+n} \in Q$ and $z_{2m+n} \in \Gamma^*$. There is an accepting computation from the last configuration $(q_{2m+n}, c^{m-n} a, z_{2m+n})$ in that sequence. Depending on n, we have two cases.

(CASE 2-2-1) if $n < m$: This case is the same as (CASE 1).

(CASE 2-2-2) if $n = m$: This case says that

$$(q_{2m}, c^m a, z_{2m}) \circlearrowleft^m (q_{3m}, a, z_{3m}) \circlearrowleft (q_{3m+1}, \epsilon, z_{3m+1})$$

for $\{q_{3m}, q_{3m+1}\} \subseteq Q$ and $\{z_{3m}, z_{3m+1}\} \subseteq \Gamma^*$. In this case, $P(w_1) = a^m b^m c^m a \in L(M, \rightarrow)$, so we can write $a^m b^m c^m a = uvxyz$, where $|vy| \geq 1$, $|vxy| \leq m$, $uv^i xy^i z \in L_2$, for all $i \geq 0$. The same argument as in the proof of Proposition 6 can be applied to reach a contradiction.

Therefore, no $\circlearrowleft$NPDA accepts L_2.   □

Since $L_{abc} \in \circlearrowleft$**NPDA** and $\{a\} \in \circlearrowleft$**NPDA**, from Theorem 4 we get that the class $\circlearrowleft$**NPDA** is not closed under concatenation. However, the reversal of L_2, that is, $\{aw \mid |w|_a = |w|_b = |w|_c\}$, is in $\circlearrowleft$**NPDA**, in fact, it is in $\circlearrowleft$**DFA**, so we get that $\circlearrowleft$**NPDA** is not closed under reversal, either.

5   Summary

We discussed three modes of tape head, however, other established modes can be considered, e.g., a two-way jumping mode. In such a mode, the tape head does not erase the letters it read. The modes can be applied to linear bounded automata, but one can show that none of these modes, including the two-way jumping mode adds to the power of linear bounded automata. Therefore, the language accepted by linear bounded automata with the mentioned alternative modes of tape head coincides with the class of context sensitive languages. Several questions remain open with respect to the jumping modes, particularly so in the case of $\circlearrowleft$. An unanswered decidability problem, which so far resisted attempts, is whether there exists an algorithm which decides $L(M, \circlearrowleft) \in$REG for a given DFA/NFA M. Here, the technique of completing $\circlearrowleft$DFA presented in Sect. 3 might help, but the problem seems rather difficult, because of the unusual languages which these classes of machines can accept.

References

1. Beier, S., Holzer, M.: Decidability of right one-way jumping finite automata. In: Hoshi, M., Seki, S. (eds.) DLT 2018. LNCS, vol. 11088, pp. 109–120. Springer, Cham (2018). https://doi.org/10.1007/978-3-319-98654-8_9
2. Beier, S., Holzer, M.: Properties of right one-way jumping finite automata. In: Konstantinidis, S., Pighizzini, G. (eds.) DCFS 2018. LNCS, vol. 10952, pp. 11–23. Springer, Cham (2018). https://doi.org/10.1007/978-3-319-94631-3_2
3. Beier, S., Holzer, M., Kutrib, M.: Operational state complexity and decidability of jumping finite automata. In: Charlier, É., Leroy, J., Rigo, M. (eds.) DLT 2017. LNCS, vol. 10396, pp. 96–108. Springer, Cham (2017). https://doi.org/10.1007/978-3-319-62809-7_6
4. Chigahara, H., Fazekas, S.Z., Yamamura, A.: One-way jumping finite automata. Int. J. Found. Comput. Sci. **27**(3), 391–405 (2016). https://doi.org/10.1142/S0129054116400165
5. Fernau, H., Paramasivan, M., Schmid, M.L.: Jumping finite automata: characterizations and complexity. In: Drewes, F. (ed.) CIAA 2015. LNCS, vol. 9223, pp. 89–101. Springer, Cham (2015). https://doi.org/10.1007/978-3-319-22360-5_8
6. Fernau, H., Paramasivan, M., Schmid, M.L., Vorel, V.: Characterization and complexity results on jumping finite automata. Theor. Comput. Sci. **679**, 31–52 (2017). https://doi.org/10.1016/j.tcs.2016.07.006
7. Hopcroft, J.E., Ullman, J.D.: Introduction to Automata Theory, Languages and Computation. Addison-Wesley, Boston (1979)
8. Kocman, R., Meduna, A.: On parallel versions of jumping finite automata. In: Janech, J., Kostolny, J., Gratkowski, T. (eds.) SDOT 2015. AISC, vol. 511, pp. 142–149. Springer, Cham (2017). https://doi.org/10.1007/978-3-319-46535-7_12
9. Krivka, Z., Meduna, A.: Jumping grammars. Int. J. Found. Comput. Sci. **26**(6), 709–732 (2015). https://doi.org/10.1142/S0129054115500409
10. Latteux, M.: Cônes rationnels commutatifs. J. Comput. Syst. Sci. **18**(3), 307–333 (1979). https://doi.org/10.1016/0022-0000(79)90039-4
11. Madejski, G.: Jumping and pumping lemmas and their applications. In: Eighth Workshop on Non-Classical Models of Automata and Applications (NCMA 2016) Short papers, pp. 25–33 (2016)
12. Meduna, A., Zemek, P.: Jumping finite automata. Int. J. Found. Comput. Sci. **23**(7), 1555–1578 (2012). https://doi.org/10.1142/S0129054112500244
13. Meduna, A., Zemek, P.: Regulated Grammars and Automata. Springer, New York (2014). https://doi.org/10.1007/978-1-4939-0369-6
14. Parikh, R.: On context-free languages. J. ACM **13**(4), 570–581 (1966). https://doi.org/10.1145/321356.321364
15. Sipser, M.: Introduction to the Theory of Computation, 2nd edn. Course Technology, Boston (2006)
16. Yu, S.: A pumping lemma for deterministic context-free languages. Inf. Process. Lett. **31**(1), 47–51 (1989). https://doi.org/10.1016/0020-0190(89)90108-7

Iterative Arrays with Self-verifying Communication Cell

Martin Kutrib[1(⊠)] and Thomas Worsch[2]

[1] Institut für Informatik, Universität Giessen, Arndtstr. 2, 35392 Giessen, Germany
kutrib@informatik.uni-giessen.de
[2] Karlsruhe Institute of Technology, Karlsruhe, Germany
worsch@kit.edu

Abstract. We study the computational capacity of self-verifying iterative arrays (SVIA). A self-verifying device is a nondeterministic device
whose nondeterminism is symmetric in the following sense. Each computation path can give one of the answers *yes*, *no*, or *do not know*. For every
input word, at least one computation path must give either the answer
yes or *no*, and the answers given must not be contradictory. It turns out
that, for any time-computable time complexity, the family of languages
accepted by SVIAs is a characterization of the so-called complementation
kernel of nondeterministic iterative array languages, that is, languages
accepted by such devices whose complementation is also accepted by such
devices. SVIAs can be sped-up by any constant multiplicative factor as
long as the result does not fall below realtime. We show that even realtime SVIA are as powerful as lineartime self-verifying cellular automata
and vice versa. So they are strictly more powerful than the deterministic devices. Closure properties and various decidability problems are
considered.

1 Introduction

One of the central questions in complexity and language theory asks for the power
of nondeterminism in bounded-resource computations. In order to gain a better
understanding of nondeterminism it has been viewed as an additional limited
resource at the disposal of time or space bounded computations. The concept
of so-called *self-verification* at least dates back to the paper [5]. It applies to
automata for decision problems and makes use of stronger notions of acceptance
and rejection of inputs.

A self-verifying device is a nondeterministic device whose nondeterminism is
symmetric in the following sense. Each computation path can give one of the
answers *yes*, *no*, or *unknown*. For every input word, at least one computation
path must give either the answer *yes* or *no*, and the answers given must not
be contradictory. So, if a computation path gives the answer *yes* or *no*, in both
cases the answer is definitely correct. This justifies the notion *self-verifying* and

A. Castillo-Ramirez and P. P. B. de Oliveira (Eds.): AUTOMATA 2019, LNCS 11525, pp. 77–90, 2019.
https://doi.org/10.1007/978-3-030-20981-0_6

is in contrast to the general case, where an answer different from *yes* does not allow to conclude whether or not the input belongs to the language. Here we study the computational capacity of self-verifying iterative arrays (SVIA).

Self-verifying finite automata have been introduced and studied in [5] and others mainly in connection with randomized Las Vegas computations. Descriptional complexity issues for self-verifying finite automata have been studied in [8]. The computational and descriptional complexity of self-verifying pushdown automata has been studied in [7]. Self-verifying cellular automata have been introduced in [12]. Some of the results in the present paper look very similar, but they require different proofs.

The paper is organized as follows. In Sect. 2 we present the basic notation and the definitions of self-verifying iterative arrays as well as an introductory example. In general, the symmetric conditions for acceptance/rejection of self-verifying devices imply immediately the effective closures of the language families accepted under complementation. In Sect. 3 this observation is turned in a characterization. Moreover, the strong speed-up by a multiplicative constant is derived for any time-computable time complexity. In Sect. 4 we explore the computational capacity of realtime SVIAs. In particular, its is shown that even realtime SVIAs are as powerful as lineartime self-verifying cellular automata and vice versa. So they are strictly more powerful than deterministic iterative arrays. Closure properties of the family of languages accepted by realtime SVIAs are studied in Sect. 5. The family is closed under the Boolean operations, reversal, concatenation, and inverse homomorphisms, while it is not closed under arbitrary homomorphisms. Finally, decidability problems are considered in Sect. 6. In particular, by a reduction of the emptiness problem it is shown that the property of being self-verifying is non-semidecidable.

2   Preliminaries and Definitions

We denote the non-negative integers by $\mathbb{N}$. Let Σ denote a finite set of letters. Then we write Σ^* for the *set of all finite words* (strings) consisting of letters from Σ. The *empty word* is denoted by λ, and $\Sigma^+ = \Sigma^* \setminus \{\lambda\}$. For the *reversal of a word* w we write w^R and $|w|$ denotes its *length*. A subset of Σ^* is called a *language* over Σ. In general, we use $\subseteq$ for *inclusions* and $\subset$ for *strict inclusions*.

A one-dimensional iterative array is a linear, semi-infinite array of finite state machines (sometimes called cells) that are identical except for the leftmost one. All but the leftmost cells are connected to their both nearest neighbors, respectively (see Fig. 1). For convenience we identify the cells by their coordinates, that is, by non-negative integers. The leftmost cell is distinguished. This so-called communication cell is connected to its right neighbor and, additionally, to the input supply which feeds the input sequentially. We assume that once the whole input is consumed an end-of-input symbol is supplied permanently. At the outset of a computation all cells are in the so-called quiescent state. The cells work synchronously at discrete time steps. Here we assume that the communication cell is a nondeterministic finite automaton while all the other cells are deterministic ones (cf. [1]). Although this is a very restricted case, for easier writing we call such devices nondeterministic.

Formally, a *nondeterministic iterative array* (NIA, for short) is a system $M = \langle S, \Sigma, F_+, s_0, \lhd, \delta_{nd}, \delta_d \rangle$, where S is the finite, nonempty set of *cell states*, Σ is the finite, nonempty set of *input symbols*, $F_+ \subseteq S$ is the set of *accepting states*, $s_0 \in S$ is the *quiescent state*, $\lhd \notin \Sigma$ is the *end-of-input symbol*, $\delta_{nd} \colon (\Sigma \cup \{\lhd\}) \times S \times S \to (2^S \setminus \emptyset)$ is the *nondeterministic local transition function for the communication cell*, $\delta_d \colon S \times S \times S \to S$ is the *deterministic local transition function for non-communication cells* satisfying $\delta_d(s_0, s_0, s_0) = s_0$.

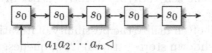

Fig. 1. Initial configuration of an iterative array.

A configuration of M at time $t \geq 0$ is a pair (w_t, c_t), where $w_t \in \Sigma^*$ is the remaining input sequence and $c_t \colon \mathbb{N} \to S$ is a mapping that maps the single cells to their current states. The initial configuration (w_0, c_0) is defined by the given input $w_0 \in \Sigma^*$ and the mapping $c_0(i) = s_0$, $i \geq 0$. Subsequent configurations are computed by the *global transition function* Δ that is induced by δ_d and δ_{nd} as follows: Let (w_t, c_t), $t \geq 0$, be a configuration. Then the set of its possible successor configurations (w_{t+1}, c_{t+1}) is defined as follows:

$$(w_{t+1}, c_{t+1}) \in \Delta((w_t, c_t)) \iff \begin{cases} c_{t+1}(0) \in \delta_{nd}(a, c_t(0), c_t(1)) \\ c_{t+1}(i) = \delta_d(c_t(i-1), c_t(i), c_t(i+1)) \end{cases}$$

for all $i \geq 1$, where $a = \lhd$, $w_{t+1} = \lambda$ if $w_t = \lambda$, and $a = a_1$, $w_{t+1} = a_2 a_3 \cdots a_n$ if $w_t = a_1 a_2 \cdots a_n$.

An input w is accepted by an NIA M if at some time step during the course of at least one computation for w the communication cell enters an accepting state. The *language accepted by M* is denoted by $L(M) = \{ w \in \Sigma^* \mid w$ is accepted by $M \}$. Let $t \colon \mathbb{N} \to \mathbb{N}$, $t(n) \geq n + 1$ be a mapping. If for each $w \in L(M)$ there is an accepting computation with at most $t(|w|)$ time steps, then M and $L(M)$ are said to be of time complexity t.

In general, the family of all languages which are accepted by some type of device X with time complexity t is denoted by $\mathscr{L}_t(X)$. If t is the function $n + 1$, acceptance is said to be in *realtime*. Since for nontrivial computations an iterative array has to read at least one end-of-input symbol, realtime has to be defined as $(n + 1)$-time. We write $\mathscr{L}_{rt}(X)$ for realtime and $\mathscr{L}_{lt}(X)$ for lineartime.

Now we turn to *self-verifying* iterative arrays (SVIA). Basically, an SVIA is an NIA, but the definition of acceptance is different. There are now three disjoint sets of states representing answers *yes*, *no*, and *neutral*. Moreover, for every input word, at least one computation path must give either the answer *yes* or *no*, and the answers given must not be contradictory. In order to implement the three possible answers the state set is partitioned into three disjoint subsets $S = F_+ \dot\cup F_- \dot\cup F_0$, where F_+ is the set of accepting states, F_- is the set of rejecting states, and $F_0 = S \setminus (F_+ \cup F_-)$ is referred to as the set of neutral states. If $M = \langle S, \Sigma, F_+, F_-, s_0, \triangleleft, \delta_{nd}, \delta_d \rangle$ is an SVIA, for each input word $w \in \Sigma^*$ and for a corresponding computation $\bar{c}$ let $S_{w,\bar{c}}$ denote the set of states entered by the communication cell during computation $\bar{c}$. For the "self-verifying property" it is required that for each $w \in \Sigma^*$ and each corresponding $\bar{c}$, $S_{w,\bar{c}} \cap F_+$ is empty if and only if $S_{w,\bar{c}} \cap F_-$ is nonempty.

If all $w \in L(M)$ are accepted and all $w \notin L(M)$ are rejected after at most $t(|w|)$ time steps, then the self-verifying iterative array M is said to be of time complexity t.

In the sequel we will often utilize the possibility of iterative arrays to simulate the data structures pushdown stores (stacks) [2,4], queues, and rings [9] without any loss of time. Here a ring is a queue that can write and erase at the same time. For pushdown stores the communication cell simulates the top of the store, for queues it simulates the front, and for rings the front and the end of the store.

We illustrate the definitions with an example.

Example 1. The nondeterministic context-free language $\{ w \in \{a,b\}^* \mid w = w^R \}$ is accepted by the SVIA $M = \langle S, \{a,b\}, F_+, F_-, s_0, \triangleleft, \delta_{nd}, \delta_d \rangle$. The basic idea is to simulate a stack whose top is the communication cell.

We set $S = (\{s_1, s_2, s_3, s_+, s_-, s_?\} \times S_{pd}) \cup \{s_0\} \cup \hat{S}_{pd}$ with $F_+ = \{s_+\} \times S_{pd}$ and $F_- = \{s_-\} \times S_{pd}$. Here S_{pd} are the register contents used by the communication cell to manage the top entries of the stack, while $\hat{S}_{pd}$ are the non-quiescent states of all but the communication cell, that realize the stack. So, the transition function δ_d just realizes the interior of the stack and is omitted here.

The idea of the construction of δ_{nd} is summarized in the following table and described below. Since we didn't make S_{pd} explicit in detail and since the state of the right neighbor of the communicating cell is only needed for updating its part of the stack, the right neighbor state is left out in the table. The current state of the communication cell is indicated as $(s_i, y^{\cdots})$ meaning that at the top of the stack is a symbol $y \in \{a,b\}$. A transition to $xy^{\cdots}$ means that x has been pushed onto, and a transition to $^{\cdots}$ means that y has been popped from the stack.

(1)	$\delta_{nd}(\triangleleft, (s_0, \bot), \_) \ni (s_+, \bot)$	accept λ
(2)	$\delta_{nd}(x, (s_0, \bot), \_) \ni (s_1, x)$	push first symbol
(3)	$\delta_{nd}(x, (s_1, y^{\cdots}), \_) \ni (s_1, xy^{\cdots})$	in s_1 continue pushing input symbols
(4)	$\delta_{nd}(x, (s_1, y^{\cdots}), \_) \ni (s_2, y^{\cdots})$	switch to s_2, dropping symbol x
(5)	$\delta_{nd}(x, (s_1, y^{\cdots}), \_) \ni (s_2, xy^{\cdots})$	switch to s_2, without dropping x
(6)	$\delta_{nd}(x, (s_2, x^{\cdots}), \_) \ni (s_2, ^{\cdots})$	continue in s_2 for matching symbols
(7)	$\delta_{nd}(\triangleleft, (s_2, \bot), \_) \ni (s_+, \bot)$	accept if everything matched
(8)	$\delta_{nd}(x, (s_2, \bar{x}^{\cdots}), \_) \ni (s_3, ^{\cdots})$	switch to s_3 if mismatch ($\bar{a} = b, \bar{b} = a$)
(9)	$\delta_{nd}(x, (s_3, y^{\cdots}), \_) \ni (s_3, ^{\cdots})$	drop remaining input symbols
(10)	$\delta_{nd}(\triangleleft, (s_3, \bot), \_) \ni (s_-, \bot)$	reject since there was a mismatch

Consider the point in time after the SVIA is in state s_2 for the first time. Let v denote the part of the input that has been pushed to the stack at that time and let u denote the part of the input that still has not been read. Then the input is vxu if (4) has been used to switch to s_2 and the input is vu if (5) was used.

If the communication cell switched from s_1 to s_2 "at the right time", that is $|u| = |v|$, then it will for the first time see input $\lhd$ and the empty stack $\bot$ simultaneously. Obviously, the decision to accept using (7) or to reject using (10) is the correct one.

If the communication cell switched from s_1 to s_2 "at the wrong time", that is $|u| \neq |v|$, then it cannot decide whether the input is a palindrome. This can be recognized by the SVIA since either all input symbols have been consumed but the stack is not empty or the stack is empty but not all input symbols have been consumed. In this case a correct behavior is obtained by entering state $s_?$ and rules that make the SVIA never leave it again.   ∎

3   Structural Properties and Speed-Up

Though we are mainly interested in fast computations, that is, realtime and lineartime computations, we allowed general time complexities in the definition of the devices (see [10] for a discussion of this general treatment of time complexity functions). However, it seems to be reasonable to consider only time complexities t that allow the communication cell to recognize the time step $t(n)$. Such functions are said to be *time-computable*. For example, the function $t(n) = n+1$ is trivially a time-computable time complexity for IAs.

Other examples are time complexities $\lfloor \frac{y}{x} \cdot n \rfloor$, for any positive integers $x < y$, polynomials $t(n) = n^k$, and exponential time complexities $t(n) = k^n$, for any integer $k \geq 2$. More details can be found in [13].

In general, the symmetric conditions for acceptance/rejection of self-verifying devices imply immediately the effective closures of the language families accepted under complementation. In order to turn this observation in a characterization, we first give evidence that self-verifying iterative arrays are in fact a generalization of deterministic iterative arrays. The proof of a corresponding result for cellular automata [12] applies here almost literally.

Lemma 2. *Any deterministic iterative array with a time-computable time complexity t can effectively be converted into an equivalent self-verifying iterative array with the same time complexity t.*

The proper inclusion $\mathscr{L}_{rt}(\text{IA}) \subset \mathscr{L}_{rt}(\text{NIA})$ is well known [1]. So, nondeterminism strengthens the computational capacity of iterative arrays. On the other hand, it is an open problem whether the family $\mathscr{L}_{rt}(\text{NIA})$ is closed under complementation. Therefore, the question whether the family $\mathscr{L}_{rt}(\text{SVIA})$ is properly included in $\mathscr{L}_{rt}(\text{NIA})$, or whether both families coincide, is of natural interest. Next we turn to relate it to the open complementation closure of $\mathscr{L}_{rt}(\text{NIA})$.

Proposition 3. *Let t be a time-computable time complexity. The family of languages $L \in \mathscr{L}_t(NIA)$, such that the complement $\overline{L}$ belongs to $\mathscr{L}_t(NIA)$ as well, coincides with the family $\mathscr{L}_t(SVIA)$.*

Proof. Given a t-time SVIA M, it is straightforward to construct an NIA that accepts the complement of $L(M)$ with the same time complexity t.

Conversely, let M_1 be an NIA accepting L and M_2 be an NIA accepting $\overline{L}$ with time complexity t. Now a t-time self-verifying iterative array M simulates M_1 and M_2 on different tracks, that is, it uses the same two channel technique of [6,14].

Then it remains to define the set of accepting states as $F_+ = \{ (s, s') \mid s \in F_1 \}$ and the set of rejecting states as $F_- = \{ (s, s') \mid s' \in F_2 \}$, where F_1 is the set of accepting states of M_1 and F_2 is the set of accepting states of M_2.   □

Proposition 3 implies that $\mathscr{L}_{rt}(SVIA)$ is properly included in $\mathscr{L}_{rt}(NIA)$ if and only if $\mathscr{L}_{rt}(NIA)$ is not closed under complementation; otherwise both families coincide.

Next, we turn to strong speed-up results for self-verifying iterative arrays from which follows that realtime is as powerful as lineartime.

Theorem 4. *Let $k \geq 1$ be a constant and t be a time complexity. Then the families $\mathscr{L}_{k \cdot t}(SVIA)$ and $\mathscr{L}_t(SVIA)$ coincide.*

Proof. A given $(k \cdot t)$-time SVIA M is simulated by a t-time SVIA M' as follows. Basically, M' performs two tasks in parallel on different tracks.

For the first task, assume that the input is fed to the communication cell of M' in k-symbol blocks, that is, k input symbols in each step. Then each k cells of M' are grouped together into one cell. In this well-known way the iterative array M' can simulate k steps of M in one step. That is, this task of M' has time complexity t and the self-verifying property, since M has time complexity $k \cdot t$ and the self-verifying property.

The second task of M' is to make the assumption for the first task true. To this end, it simulates a ring store whose front (and end) is the communication cell. Now the communication cell starts to guess k input symbols in every step. These symbols are fed to the first task. Additionally, the communication step guesses when the end-of-input symbol appears. From that time step on no further input symbols are guessed. In order to verify that the guesses are correct, the k symbols are entered at the end of the ring store respectively. In each step, the symbol at the front of the ring is removed and compared with the actual input symbol. If both match, the guessed symbol is correct, otherwise it is not. In case of a mismatch or a wrongly guessed number of input symbols the second task remains in a neutral state. If it has guessed the input correctly, it enters a positive state.

Finally, M' accepts if and only if the second task guesses the input correctly and the first task accepts. That is, if the actual input is accepted by M. The iterative array M' rejects if and only if the second task guesses the input correctly and the first task rejects. That is, if the actual input is rejected by M. So, M' has the self-verifying property.   □

Corollary 5. *The families $\mathscr{L}_{rt}(SVIA)$ and $\mathscr{L}_{lt}(SVIA)$ coincide.*

4   Computational Capacity

The first question in connection with the computational capacity of realtime SVIA is the impact of the (restricted) nondeterminism. Does it increase the capacity? More precisely, we are interested in the question whether the computing power of realtime SVIA is strictly stronger than that of realtime IA.

Example 1 shows that the mirror language is accepted by a realtime SVIA. However, by using a completely different algorithm the language is accepted by some deterministic realtime IA as well [3]. So, it cannot be used as a witness for the strictness of the inclusion $\mathscr{L}_{rt}(\text{IA}) \subset \mathscr{L}_{rt}(\text{SVIA})$. Nevertheless, the strictness follows from a more general result below and is stated in Proposition 10.

Corollary 6. *The family of languages accepted by self-verifying pushdown automata is strictly included in the family $\mathscr{L}_{rt}(SVIA)$.*

In order to discuss further comparisons we now turn to results that show the strong computational capacity of realtime SVIA.

While iterative arrays fetch their input sequentially through the communication cell, so-called cellular automata obey a parallel input mode. In a preinitial step their cells fetch an input symbol. That is, there are as many cells as input symbols. So, a two-way cellular automaton (CA) is a linear array of identical finite automata which are numbered $1, 2, \ldots, n$. Except for border cells the state transition depends on the current state of a cell itself and those of its both nearest neighbors. Border cells receive a boundary symbol # on their free input lines. An input w is accepted by a cellular automaton if at some time step during some computation the leftmost cell enters an accepting state. For cellular automata, realtime is defined to be $t(n) = n$. Cellular automata whose first state transitions are nondeterministic, whose further transitions are deterministic, and that have the self-verifying property (SVCA) are studied in [12].

Lemma 7. *The family $\mathscr{L}_{lt}(SVIA)$ is included in $\mathscr{L}_{rt}(SVCA)$.*

Lemma 7 gives an upper bound for languages accepted by lineartime SVIA. It shows that a sequential input mode and one nondeterministic cell can be traded for parallel input mode and all cells nondeterministic once only. In fact, the upper bound is sharp, that is, the converse is also true.

Lemma 8. *The family $\mathscr{L}_{lt}(SVCA)$ is included in $\mathscr{L}_{rt}(SVIA)$.*

Proof. The possibility to speed-up SVCAs by a constant factor is shown in [12]. That is, $\mathscr{L}_{lt}(\text{SVCA}) = \mathscr{L}_{rt}(\text{SVCA})$. So, given some realtime SVCA M with state set S and set of input symbols Σ, we construct an equivalent realtime SVIA M' as follows. Basically, M' works in three phases. First it guesses and generates a configuration that represents the fourfold packed initial configuration of M. Then this packed part is synchronized. Finally, the synchronized cells simulate M

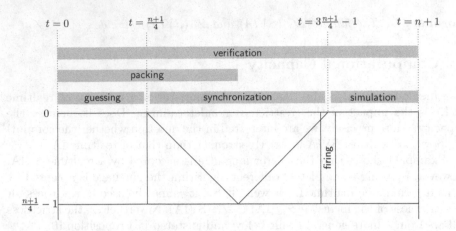

Fig. 2. Simulation phases.

whereby four steps are simulated in one step. The phases are depicted in Fig. 2. In parallel, the guesses are verified.

Phase 1: Let n denote the length of the input. In the following we assume that $n+1$ is a multiple of four. The generalization of the simulation to the other cases is a straightforward adaption.

So, during the first $\frac{n+1}{4}$ time steps the communication cell of M' guesses four input symbols in every step. Additionally, the communication cell guesses a mapping $(\Sigma \cup \{\#\}) \times \{a\} \times (\Sigma \cup \{\#\}) \to S$ for each (guessed) input symbol $a \in \Sigma$. These mappings are used for the simulation of the nondeterministic first transitions of the cells of M. The blocks of four input symbols together with the corresponding mappings are shifted to the right such that each of the leftmost $\frac{n+1}{4}$ cells gets one block of symbols and mappings. When the communication cell guesses the end-of-input symbol the first phase ends.

In order to verify that the guesses are correct, the four symbols are entered at the end of a ring store respectively. In each step, the symbol at the front of the ring is removed and compared with the actual input symbol. If both match, the guessed symbol is correct, otherwise it is not. In case of a mismatch or a wrongly guessed number of input symbols the computation blocks in a neutral state.

Phase 2: At time step $\frac{n+1}{4}+1$ the communication cell initiates an FSSP synchronization of the leftmost $\frac{n}{4}$ cells. The blocks of four input symbols together with the corresponding mappings arrive at their destination cells $0 \le i \le \frac{n+1}{4}-1$ at time $2i+1$. The initial signal for the FSSP arrives at cell i at time $\frac{n+1}{4}+1+i > 2i+1$. So, each cell starts Phase 2 after finishing Phase 1. Altogether, Phase 2 is finished when all cells are synchronized at time $\frac{n+1}{4}+1+2 \cdot \frac{n+1}{4}-2 = 3 \cdot \frac{n+1}{4}-1$.

Phase 3: Due to the compressed representation, M' can simulate M with fourfold speed. In order to simulate the nondeterministic transitions which the cells of M perform during the first time step, the cells of M' apply the nondeterministically guessed mappings to its local configurations. Thus, M' simulates the nth step of M at time step $3 \cdot \frac{n+1}{4} - 1 + \frac{n+1}{4} = n$.

Since the verification of the guessed input takes $n+1$ time steps, we conclude that the total time complexity of M' is $t(n) = n + 1$, that is realtime.

Finally, M' accepts if and only if the input has been guessed correctly and M accepts, and it rejects if and only the input has been guessed correctly and M rejects. Since M has the self-verifying property, M' is self-verifying as well. So, we have $L(M') = L(M)$. $\qquad\square$

Lemmas 7 and 8 reveal the equality of the next theorem.

Theorem 9. $\mathscr{L}_{lt}(SVCA) = \mathscr{L}_{rt}(SVCA) = \mathscr{L}_{rt}(SVIA) = \mathscr{L}_{lt}(SVIA)$.

In particular, now we can deduce that even the restricted nondeterminism gained in considering a self-verifying communication cell strictly increases the computational capacity of realtime iterative arrays. That is, the inclusion $\mathscr{L}_{rt}(\text{IA}) \subset \mathscr{L}_{rt}(\text{SVIA})$ is strict.

Proposition 10. *The family $\mathscr{L}_{rt}(IA)$ is strictly included in $\mathscr{L}_{rt}(SVIA)$.*

Proof. The relations $\mathscr{L}_{rt}(\text{IA}) \subset \mathscr{L}_{lt}(\text{IA}) = \mathscr{L}_{lt}(\text{CA})$ are known (see, for example, [10]). Since $\mathscr{L}_{lt}(\text{IA}) \subseteq \mathscr{L}_{lt}(\text{SVIA})$ the assertion follows. $\qquad\square$

Theorem 9 shows that a sequential input mode and one nondeterministic cell can be traded for parallel input mode and all cells nondeterministic once only, and vice versa. To this end, it does not matter whether the computations are in realtime or lineartime. But what about the world beyond lineartime? Are self-verifying arrays stronger than deterministic ones? Or weaker than nondeterministic ones? The open question of the strictness of one of the inclusions $\mathscr{L}_{lt}(\text{IA}) \subseteq \mathscr{L}_{rt}(\text{SVIA}) \subseteq \mathscr{L}(\text{SVCA}) \subseteq \mathscr{L}(\text{NCA})$ is strongly related to famous open problems in complexity theory (see [11]). Note that at the top of this hierarchy are devices that may have an exponential time complexity (due to the space bound).

5 Closure Properties

Here we turn to explore the closure properties of the family of realtime SVIA languages. They are summarized in Table 1. We start with the Boolean operations.

Proposition 11. *The family of languages accepted by realtime SVIA is closed under complementation, union, and intersection.*

The closure under reversal is of crucial importance. It is an open problem for $\mathscr{L}_{rt}(\text{CA})$ and, equivalently, for $\mathscr{L}_{lt}(\text{OCA})$ (OCAs are CAs where information can only be passed from right to left, that is, the new state of a cell does not depend on that of its left neighbor). Moreover, it is linked with the open closure property under concatenation for the same family and, hence, with the question whether lineartime CAs are more powerful than realtime CAs. It is known that the family $\mathscr{L}_{rt}(\text{IA})$ is not closed under reversal, while the family $\mathscr{L}_{lt}(\text{IA})$ is closed.

Proposition 12. *The family of languages accepted by realtime SVIA is closed under reversal.*

Proof. Given some realtime SVIA M, Theorem 9 says that there is an equivalent lineartime SVCA M' with transition functions δ_{nd} and δ_d.

Now, the arguments of the local transition functions are interchanged. That is, $\delta'_d(s_3, s_2, s_1)$ is defined to be $\delta_d(s_1, s_2, s_3)$, and $\delta'_{nd}(s_3, s_2, s_1)$ is defined to be $\delta_{nd}(s_1, s_2, s_3)$. The resulting device M'' with transition functions δ'_{nd} and δ'_d. accepts the reversal $L(M)^R$ at the *rightmost cell*. Then the result is sent as a signal to the leftmost cell. Altogether, M'' is still a lineartime SVCA and, thus, $L(M)^R \in \mathscr{L}_{lt}(\text{SVCA}) = \mathscr{L}_{rt}(\text{SVIA})$. □

Proposition 13. *The family of languages accepted by realtime SVIA is closed under concatenation.*

Proof. Let $L_1, L_2 \in \mathscr{L}_{rt}(\text{SVIA})$. If the empty word belongs to L_1 then language L_2 belongs to the concatenation and vice versa. Since the family of languages accepted by realtime SVIA is closed under union, it remains to consider languages $L_1, L_2 \in \mathscr{L}_{rt}(\text{SVIA})$ that do not contain the empty word. Let M_1 and M_2 be realtime SVIA that accept L_1 and L_2.

Since the family $\mathscr{L}_{rt}(\text{SVIA})$ is closed under reversal, there is a realtime SVIA M_2^R that accepts the reversal L_2^R of L_2.

A realtime SVIA M that accepts the concatenation $L_1 \cdot L_2$ works as follows. First we describe two tasks that are performed by M in parallel.

Basically, the first task is to read the input and to simulate M_1. In addition, the input is stored into a ring whose front is the communication cell. Moreover, in any simulation step, M tests whether it would accept or reject the input prefix read so far by checking if it would accept or reject when the next input symbol were the end-of-input symbol $\lhd$. If the current input prefix is accepted or rejected, the input symbol stored into the ring is marked suitably.

The second task is to guess the reversal of the input symbol by symbol. The guessed reversal is stored into a pushdown store whose top is the communication cell. Additionally, the realtime SVIA M_2^R is simulated on the guessed input. Similarly as for the first task, if the current prefix of the guessed input would be accepted or rejected, the guessed input symbol stored into the pushdown store is marked suitably.

Let $x_1 x_2 \cdots x_n$ be the actual input. It is stored in the ring when the end-of-input symbol appears. At that time, let $y_1 y_2 \cdot y_n$ be the content of the pushdown store (from top to bottom). Clearly, M has guessed the reversal of the input correctly if and only if $x_1 x_2 \cdots x_n = y_1 y_2 \cdots y_n$. So, after having read the end-of-input symbol, the SVIA M verifies the guessed reversal of the input by successively removing symbols from the ring and pushdown store and testing whether they match. If M detects any mismatch it blocks in a neutral state.

Now, assume that the reversal of the input has been guessed correctly.

Already while verifying the guesses by successively scanning the ring and the queue, the SVIA M tests whether for some $1 \leq i \leq n-1$ input symbol x_i is marked by the simulation of M_1 and symbol y_{i+1} is marked by the simulation of M_2^R.

Case 1: $x_1 \cdots x_n \in L_1 L_2$, say $x_1 \cdots x_i \in L_1$. Then there are computations by M_1 accepting $x_1 \cdots x_i$ and by M_2^R accepting $y_n y_{n-1} \cdots y_{i+1} = x_n x_{n-1} \cdots x_{i+1}$ $= (x_{i+1} \cdots x_n)^R$. Having M accept an input iff x_i is marked by M_1 and symbol y_{i+1} is marked by M_2^R makes M accept all words in $L_1 L_2$.

Case 2: $x_1 \cdots x_n \notin L_1 L_2$. In this case for each i either $x_1 \cdots x_i \notin L_1$ or $x_n \cdots x_{i+1} \notin L_2^R$ or both. That means that for each i there are computations by M_1 and M_2^R for the respective inputs such that at least one of both rejects. Hence, there will be a computation for M which correctly explicitly rejects an input, if for any two adjacent cells always at least one of them is marked rejecting.

In any other case, the leftmost cell remains in a neutral state.

For the computation on input of length n the SVIA M takes $n+1$ steps to read (and guess) the input for the tasks, and further $n+1$ steps to verify the guesses and test the markings. So, M works in lineartime which ca be sped-up to realtime.   □

Next, we turn to the operations homomorphism and inverse homomorphism.

Proposition 14. *The family of languages accepted by realtime SVIA is not closed under arbitrary homomorphisms.*

Proof. It is shown in [12] that every recursively enumerable language would be contained in the family $\mathscr{L}_{rt}(\text{SVOCA})$ if the family were closed under arbitrary homomorphisms. Since $\mathscr{L}_{rt}(\text{SVOCA}) \subsetneq \mathscr{L}_{rt}(\text{SVCA}) = \mathscr{L}_{rt}(\text{SVIA})$ the same is true for the family $\mathscr{L}_{rt}(\text{SVIA})$, a contradiction due to the time bound.   □

Proposition 15. *The family of languages accepted by realtime SVIA is closed under inverse homomorphisms.*

The closure properties of $\mathscr{L}_{rt}(\text{SVIA})$ with respect to iteration (Kleene star) and non-erasing homomorphisms are open problems. They are settled for non-deterministic devices since, basically, for iteration it is sufficient to guess the positions in the input at which words are concatenated, and for non-erasing homomorphism it is sufficient to guess the pre-image of the input. However, self-verifying devices have to reject explicitly if the input does not belong to the language. Intuitively, this means that they have to 'know' that all possible guesses either do not lead to accepting computations or are 'wrong.'

6   Decidability Questions

First we note that the membership problem is obviously decidable for SVIAs obeying a time-computable time complexity.

On the other hand, in [10] it is observed that for any language family that effectively contains $\mathscr{L}_{rt}(\text{IA})$, the problems emptiness, universality, finiteness, infiniteness, regularity, and context-freeness are not semidecidable. Since we know $\mathscr{L}_{rt}(\text{IA}) \subset \mathscr{L}_{lt}(\text{IA}) \subseteq \mathscr{L}_{lt}(\text{SVIA}) = \mathscr{L}_{rt}(\text{SVIA})$ we derive the next corollary.

Table 1. Closure properties of the language family $\mathscr{L}_{rt}(\mathrm{SVIA})$ in comparison with the families $\mathscr{L}_{rt}(\mathrm{IA})$ and $\mathscr{L}_{lt}(\mathrm{IA})$, where h_λ denotes λ-free homomorphisms.

Family	$-$	$\cup$	$\cap$	R	$\cdot$	$*$	h_λ	h	h^{-1}
$\mathscr{L}_{rt}(\mathrm{SVIA})$	✓	✓	✓	✓	✓	?	?	✗	✓
$\mathscr{L}_{rt}(\mathrm{IA})$	✓	✓	✓	✗	✗	✗	✗	✗	✓
$\mathscr{L}_{lt}(\mathrm{IA})$	✓	✓	✓	✓	?	?	?	✗	✓

Corollary 16. *The problems emptiness, universality, finiteness, infiniteness, inclusion, equivalence, regularity, and context-freeness are not semidecidable for realtime IAs and thus for realtime SVIAs.*

In [12] it is shown that the problem to decide whether a given realtime one-way cellular automaton is self-verifying or not is undecidable. Unfortunately, the result has no direct implications for the same question for iterative arrays. However, the undecidability for cellular automata is shown by a reduction of the emptiness problem. We turn to prove the undecidability for iterative arrays as well. Moreover, we use a reduction of the emptiness and universality problem, but the reduction itself is different. Since general iterative arrays do not have neutral or rejecting states (only accepting and non-accepting states), there is no partitioning of the state set. So, the decidability can be asked for a given fixed partitioning or for the existence of a partitioning. We first consider the latter question.

Theorem 17. *Given a realtime (non)deterministic iterative array M with state set S and accepting states F_+, it is not even semidecidable whether there exists $F_- \subseteq (S \setminus F_+)$ such that M is an SVIA with respect to the sets F_+ and F_-.*

Proof. Let $M_0 = \langle S, \Sigma, F_+, s_0, \lhd, \delta_{nd}, \delta_d \rangle$ be an arbitrary realtime IA. We safely may assume that a cell which has left the quiescent state will never enter the quiescent state again. This behavior can be implemented by adding a new state that plays the role of the quiescent state. If necessary, the new state can be entered instead of s_0.

We modify M_0 to $M_1 = \langle S', \Sigma', F'_+, s_0, \lhd, \delta'_{nd}, \delta'_d \rangle$ by adding a new input symbol $\$$ and two new states p_+ and p_0. So, we set $S' = S \cup \{p_+, p_0\}$, $\Sigma' = \Sigma \cup \{\$\}$, and $F'_+ = F_+ \cup \{p_+\}$. The intention is that a $\$$ in the input causes the IA to simulate a step on the end-of-input symbol $\lhd$ (in restricted form) and to reinitialize the computation by letting the cells enter the quiescent state again (which is impossible in M_1). Therefore, the transition function δ'_{nd} is basically δ_{nd} extended by transitions for the input symbol $\$$ and the states p_+ and p_0. When a $\$$ appears in the input, the communication cell enters state p_+ if it could enter an accepting state on the end-of-input symbol $\lhd$. For all $s_1, s_2 \in S$,

$$\delta'_{nd}(\$, s_1, s_2) = \{p_+\} \text{ if } \delta_{nd}(\lhd, s_1, s_2) \cap F_+ \neq \emptyset.$$

Otherwise it enters state p_0: $\delta'_{nd}(\$, s_1, s_2) = \{p_0\}$ if $\delta_{nd}(\lhd, s_1, s_2) \cap F_+ = \emptyset$. In state p_+ or p_0 the computation continues as it would from the very beginning.

For $p \in \{p_+, p_0\}$, all $a \in \Sigma' \cup \{\lhd\}$, and all $s \in S'$, $\delta'_{nd}(a, p, s) = \delta'_{nd}(a, s_0, s_0)$. In order to implement the reinitialization of the other cells, recall that δ_d drives no non-quiescent cell into the quiescent state. So, we can utilize the quiescent state as a signal sent by the communication cell. The signal causes the reinitialization of the cells passed through. So, the transition function δ'_d is basically δ_d extended as follows. For $p \in \{p_+, p_0\}$ and all $s_1, s_2 \in S$,

$$\delta'_d(p, s_1, s_2) = s_0, \quad \delta'_d(s_0, s_1, s_2) = s_0, \quad \delta'_d(s_1, s_0, s_2) = \delta(s_1, s_0, s_0).$$

Therefore $L(M_1)$ consists of all concatenations of \$ separated words u_i such that at least one u_i is in $L(M_0)$. In particular $L(M_1) \cap \Sigma^* = L(M_0)$.

We claim that there exists $F'_- \subseteq (S' \setminus F'_+)$ such that the iterative array $M_2 = \langle S', \Sigma', F'_+, F'_-, s_0, \lhd, \delta'_{nd}, \delta'_d \rangle$ is self-verifying if and only if $L(M_0)$ is empty or coincides with Σ^*. Observe that $L(M_2) = L(M_1)$ because both have the same set of accepting states.

If $L(M_0)$ is empty then $L(M_1)$ is empty. Therefore, the communication cell will never enter an accepting state from F'_+. So, we safely may set $F'_- = (S' \setminus F'_+)$ and obtain that M_2 is self-verifying. Similarly, if $L(M_0) = \Sigma^*$ then $L(M_1) = \Sigma'^*$, and we safely may set $F'_- = \emptyset$ to obtain a self-verifying IA.

Now assume that $L(M_0)$ and, thus, $L(M_2)$ neither be empty nor contain all words over the input alphabet. Then there exists some $u \in L(M_0)$ and some $v \notin L(M_0)$. We consider the computation of M_2 on input $u\$v$. Since M_0 accepts u, the IA M_2 enters an accepting state while processing the input prefix $u\$$ (its computation is a simulation of M_0 on $u\lhd$). Then the computation of M_2 is reinitialized and continues with a simulation of M_0 on input v. Since $v \notin L(M_0)$, in this phase, M_2 cannot accept v either. However, since it already was in an accepting state and its overall answer is already yes, M_2 cannot enter a contradictory rejecting state in this phase either. This implies that the communication cell of M_2 on input v will only assume neutral states and, thus, neither accept nor reject v. That is, M_2 is not self-verifying and the claim follows.

From the construction of M_2 and the claim we conclude that the semidecidability of the problem in question implies the semidecidability of the emptiness or universality problem for realtime IAs contradicting Corollary 16.   ⊓

What about the undecidability if we provide a partitioning of its state set? Can we test if this partitioning makes the IA self-verifying? The answer is no, since for a given realtime iterative array with accepting state set F_+ there are only finitely many partitions induced by setting $F_- \subseteq (S \setminus F_+)$. All these could be tested in parallel. Now the problem in question can be semidecided if the test is successful for at least one partitioning.

Corollary 18. *Given a realtime (non)deterministic iterative array M with state set S and partitioning $S = F_+ \,\dot\cup\, F_- \,\dot\cup\, F_0$, it is not semidecidable whether M is an SVIA with respect to the partitioning.*

By Lemma 2 any deterministic iterative with a time-computable time complexity can effectively be made self-verifying. But it is non-semidecidable whether it already *is* self-verifying. This non-semidecidability carries immediately over to nondeterministic iterative arrays. However, it is an open problem whether any nondeterministic iterative with a time-computable time complexity can effectively be made self-verifying. In fact, it is an open problem whether the family of languages accepted by realtime nondeterministic iterative arrays is closed under complementation or not.

References

1. Buchholz, Th., Klein, A., Kutrib, M.: Iterative arrays with limited nondeterministic communication cell. In: Words, Languages and Combinatorics III, pp. 73–87. World Scientific Publishing (2003)
2. Buchholz, Th., Kutrib, M.: Some relations between massively parallel arrays. Parallel Comput. **23**, 1643–1662 (1997)
3. Cole, S.N.: Real-time computation by n-dimensional iterative arrays of finite-state machines. IEEE Trans. Comput. **C−18**, 349–365 (1969)
4. Čulik II, K., Yu, S.: Iterative tree automata. Theoret. Comput. Sci. **32**, 227–247 (1984)
5. Ďuriš, P., Hromkovič, J., Rolim, J.D.P., Schnitger, G.: Las Vegas versus determinism for one-way communication complexity, finite automata, and polynomial-time computations. In: Reischuk, R., Morvan, M. (eds.) STACS 1997. LNCS, vol. 1200, pp. 117–128. Springer, Heidelberg (1997). https://doi.org/10.1007/BFb0023453
6. Dyer, C.R.: One-way bounded cellular automata. Inf. Control **44**, 261–281 (1980)
7. Fernau, H., Kutrib, M., Wendlandt, M.: Self-verifying pushdown automata. In: Non-Classical Models of Automata and Applications (NCMA 2017). books@ocg.at, vol. 329, pp. 103–117. Austrian Computer Society, Vienna (2017)
8. Jirásková, G., Pighizzini, G.: Optimal simulation of self-verifying automata by deterministic automata. Inf. Comput. **209**, 528–535 (2011)
9. Kutrib, M.: Cellular automata - a computational point of view. In: Bel-Enguix, G., Jiménez-López, M.D., Martín-Vide, C. (eds.) New Developments in Formal Languages and Applications. SCI, vol. 113, pp. 183–227. Springer, Heidelberg (2008). https://doi.org/10.1007/978-3-540-78291-9_6
10. Kutrib, M.: Cellular automata and language theory. In: Meyers, R. (ed.) Encyclopedia of Complexity and System Science, pp. 800–823. Springer, New York (2009). https://doi.org/10.1007/978-0-387-30440-3
11. Kutrib, M.: Complexity of one-way cellular automata. In: Isokawa, T., Imai, K., Matsui, N., Peper, F., Umeo, H. (eds.) AUTOMATA 2014. LNCS, vol. 8996, pp. 3–18. Springer, Cham (2015). https://doi.org/10.1007/978-3-319-18812-6_1
12. Kutrib, M., Worsch, T.: Self-verifying cellular automata. In: Mauri, G., El Yacoubi, S., Dennunzio, A., Nishinari, K., Manzoni, L. (eds.) ACRI 2018. LNCS, vol. 11115, pp. 340–351. Springer, Cham (2018). https://doi.org/10.1007/978-3-319-99813-8_31
13. Mazoyer, J., Terrier, V.: Signals in one-dimensional cellular automata. Theoret. Comput. Sci. **217**, 53–80 (1999)
14. Smith III, A.R.: Real-time language recognition by one-dimensional cellular automata. J. Comput. Syst. Sci. **6**, 233–253 (1972)

Generic Properties in Some Classes
of Automaton Groups

Thibault Godin[1,2]($\boxtimes$)

[1] IECL, UMR 7502 CNRS and Université de Lorraine, Nancy, France
[2] IMAG, UMR 5149 CNRS and Université de Montpellier, Montpellier, France
thibault.godin@umontpellier.fr

Abstract. We prove, for various important classes of Mealy automata, that almost all generated groups have an element of infinite order. In certain cases, we also prove that some other properties, such as exponential growth, are generic.

1 Introduction

The class of groups generated by Mealy automata presents a considerable variety of behaviours and has been widely used since the eighties as a powerful source of interesting groups [5,14,15,26]. It seems natural to try to produce new examples of groups to be studied by picking a random Mealy automaton and considering the group it generates, or to try to get an interesting distribution over some class of groups starting from a distribution over some class of Mealy automata [12]. This approach also raises a natural question: *"what does a typical automaton group look like?"*. In this paper, we tackle this problem and give partial answers for several important and well-studied classes, by proving that automata belonging to the class of reversible, reset, or polynomial activity automata generate with great probability a group having at least one element of infinite order. In particular, it means that these groups are generically infinite and not Burnside.

Another motivation for this paper is that the ORDER PROBLEM—how to decide whether an element generates an infinite group—was recently proven undecidable among automaton groups [4,11], while the FINITENESS PROBLEM—how to decide whether the whole group is infinite—is known to be undecidable for automaton semigroups but remains open for automaton groups [7,10]. On the other hand, some classes of automaton (semi)groups are known to have decidable ORDER PROBLEM [3,6]. Our results provide probabilistic answers for these problems.

Depending on the class, we also get stronger or additional statements, among others, the groups generated by reversible or reset Mealy automata have generically exponential growth.

© IFIP International Federation for Information Processing 2019
Published by Springer Nature Switzerland AG 2019
A. Castillo-Ramirez and P. P. B. de Oliveira (Eds.): AUTOMATA 2019, LNCS 11525, pp. 91–103, 2019.
https://doi.org/10.1007/978-3-030-20981-0_7

The proposed proofs vary strongly with the considered class and rely on the structural properties of the automata. In particular, the case of general invertible Mealy automata remains open.

In order to simplify the statements, we will use the informal "let $\mathcal{A}$ be a random automaton in $\mathcal{C}$" instead of the formal "let $\mathcal{A}$ be a random variable uniformly distributed over the set $\mathcal{C}$". All probabilistic statements should be understood accordingly.

2   Automaton Groups

We recall that the *order* of an element g of a group G is the least (strictly positive) integer α such that $g^\alpha = \mathbb{1}$. If such an integer does not exist, we say that g has infinite order. Equivalently, the order of g is the cardinal of the subgroup it generates, hence having an element of infinite order implies the infiniteness of the whole group.

If X is a finite set then X^k denotes the set of words of length k, and X^* (*resp.* X^+) the set of words of arbitrary (*resp.* positive) length. We take as a convention that elements of X^ℓ, $\ell > 1$ are represented with a bold font.

2.1   Mealy Automata and Automaton (Semi)Groups

A *Mealy automaton* is a 4-tuple $\mathcal{A} = (Q, \Sigma, \delta, \rho)$ where Q and Σ are finite sets, called the *stateset* and the *alphabet* respectively, $\delta = \{\delta_x : Q \to Q\}_{x \in \Sigma}$ is a set of functions called *transition* functions, and $\rho = \{\rho_q : \Sigma \to \Sigma\}_{q \in Q}$ is a set of functions called *production* functions. Examples of such automata are presented on Fig. 1, and we refer the reader to [17] for a more complete introduction.

The map ρ_q extends to a length-preserving map on words $\rho_q : \Sigma^* \to \Sigma^*$ by the recursive definition:

$$\forall i \in \Sigma, \ \forall \mathbf{s} \in \Sigma^*, \qquad \rho_q(i\mathbf{s}) = \rho_q(i)\rho_{\delta_i(q)}(\mathbf{s}) \, .$$

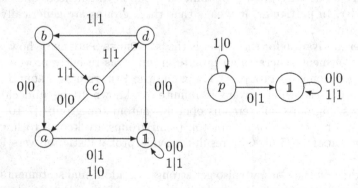

Fig. 1. The automaton generating the Grigorchuk group (left) and the adding machine, generating $\mathbb{Z}$ (right). The Grigorchuk automaton has stateset $\{a, b, c, d, \mathbb{1}\}$ and the adding machine $\{p, \mathbb{1}\}$. Both automata have alphabet $\{0, 1\}$

We can also extend the set of maps ρ to words of states $\mathbf{u} \in Q^*$ by composing the production functions associated with the letters of $\mathbf{u}$:

$$\forall q \in Q, \ \forall \mathbf{u} \in Q^*, \qquad \rho_{q\mathbf{u}} = \rho_{\mathbf{u}} \circ \rho_q .$$

Likewise, we extend the functions δ to words of state and words via

$$\forall i \in \Sigma, \ \forall \mathbf{s} \in \Sigma^*, \ \forall q \in Q, \ \forall \mathbf{u} \in Q^*, \quad \delta_i(q\mathbf{u}) = \delta_i(q)\delta_{\rho_q(i)}(\mathbf{u}) \text{ and } \delta_{i\mathbf{s}} = \delta_i \circ \delta_{\mathbf{s}} .$$

For each automaton transition $q \xrightarrow{x|\rho_q(x)} \delta_x(q)$, we associate the *cross-transition* depicted in (cross):

$$q \ \underset{\rho_q(x)}{\overset{x}{+}} \ \delta_x(q) \qquad\qquad \text{(cross)}$$

The production functions $\rho_q : \Sigma^* \to \Sigma^*$ of an automaton $\mathcal{A}$ generate a semigroup $\langle \mathcal{A} \rangle_+ = \{\rho_{\mathbf{u}} : \Sigma^* \to \Sigma^* | \mathbf{u} \in Q^+\}$.

A Mealy automaton is *invertible* when the functions ρ are permutations of Σ. When a Mealy automaton is invertible one can define its *inverse* $\mathcal{A}^{-1}$ by

$$p \xrightarrow{x|y} q \in \mathcal{A} \Leftrightarrow p^{-1} \xrightarrow{y|x} q^{-1} \in \mathcal{A}^{-1} .$$

Whenever a Mealy automaton is invertible we can consider the *group* $\langle \mathcal{A} \rangle$ it generates:

$$\langle \mathcal{A} \rangle = \langle \rho_q \mid q \in Q \rangle = \{\rho_{\mathbf{u}}^{\pm 1} \mid \mathbf{u} \in Q^*\} .$$

A group (*resp.* a semigroup) is an *automaton group* (*resp. semigroup*) if it can be generated by some Mealy automaton.

Given a Mealy automaton $\mathcal{A} = (Q, \Sigma, \delta, \rho)$, its *dual* is the Mealy automaton $\partial\mathcal{A} = (\Sigma, Q, \rho, \delta)$ where the roles of the stateset and of the alphabet are exchanged. Its ℓ-th power is the automaton $(Q^\ell, \Sigma, \delta, \rho)$ where the production and transition functions have been naturally extended. We define also the automaton $^\ell\mathcal{A} = (Q, \Sigma^\ell, \delta, \rho) = \partial(\partial\mathcal{A})^\ell$ acting on sequences of ℓ letters and remark that this operation does not change the generated semigroup, *i.e.* $\langle ^\ell\mathcal{A} \rangle_+ = \langle \mathcal{A} \rangle_+$.

An automaton is called *minimal* if each state induces a different element in the generated group. Given an automaton $\mathcal{A}$, one can algorithmically construct its minimization $\mathsf{m}\mathcal{A}$, which generates the same group [2].

From an algebraic point of view, it is convenient to describe the elements of an automaton group via the so-called wreath recursions. For any g in an automaton group $\langle \mathcal{A} \rangle$ on alphabet $\Sigma = \{1, \ldots, k\}$ and any word $\mathbf{s} \in \Sigma^*$, let $g \cdot \mathbf{s}$ denotes the image of $\mathbf{s}$ by g, and $g_{|\mathbf{s}}$ the unique $h \in \langle \mathcal{A} \rangle$ satisfying $g \cdot (\mathbf{st}) = (g \cdot \mathbf{s})h \cdot \mathbf{t}$ for all $\mathbf{t} \in \Sigma^*$. The *wreath recursion* of g is:

$$g = (g_{|1}, \ldots, g_{|x_k})\sigma_g,$$

where $\sigma_g \in S_k$ denotes the permutation on Σ induced by g.

In what follows, $\mathcal{A}$ will denote, if not explicited, an invertible Mealy automaton $(Q, \Sigma, \delta, \rho)$.

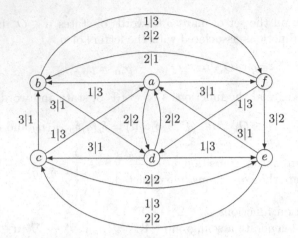

Fig. 2. A 3-letter 6-state invertible reversible non-bireversible Mealy automaton.

2.2 Classes of Mealy Automata

We now describe several important classes of (invertible) Mealy automata. Since invertibility is required to generate a group, it will be assumed throughout the paper.

An automaton $\mathcal{A} = (Q, \Sigma, \delta, \rho)$ is called *reversible* when the functions δ are permutations of Q. If an automaton is invertible then its dual is reversible. A Mealy automaton is *bireversible* if both itself and its inverse are invertible and reversible.

Another, somewhat opposite, restriction on the transition function leads to the class of *reset* automata, studied *e.g.* in [7,24]. An automaton $\mathcal{A}$ is called *reset* if there exists a function $\phi : \Sigma \to Q$ such that $\forall x, \forall q, \delta_x(q) = \phi(x)$. In other words, all the arrows labelled by an input letter x lead to the same state $\phi(x)$. Up to renaming the states and pruning the automaton of its vertices without ingoing edges (which does not change the finiteness of the generated group nor the existence of element of infinite order), we may assume that all studied reset automata are *unfolded*, *i.e.* that $Q = \Sigma$ and $\phi = \mathbb{1}$.

Another class of Mealy automata linked to the cycle structure is defined in [23], via the *activity*. Assume that there is a unique state inducing the identity in the group, denoted $\mathbb{1}$. The activity of an automorphism $t \in \langle \mathcal{A} \rangle$ is defined as the function
$$\alpha_t : \ell \mapsto \left| \{x \in \Sigma^\ell, t_{|x} \neq \mathbb{1}\} \right|.$$

It is known that the activity α_t is polynomial if and only if there is not two non-trivial simple cycles accessible one from another in the automaton $\mathcal{A}$, and that in this case the degree of the polynomial is the maximal number of nontrivial cycles that can be reached along of a simple cycle minus one. For a fixed alphabet Σ, we denote Pol (*resp.* Pol(d)) the set of all Mealy automata with polynomial activity (*resp.* with activity bounded by a polynomial of degree d), and in particular we call *bounded* (*resp. finitary*) the set $\mathrm{Pol}(0)$ (*resp.* $\mathrm{Pol}(-1) = \{t, \alpha_t(\ell) \to 0\}$). For instance both automata on Fig. 1 have bounded activity. Notice that every finitary automaton generates a finite group [22].

In [6], a tool is defined to understand the orbits of elements of Σ^* under the action of an automorphism described by a Mealy automaton. Given $\mathcal{A} = (Q, \Sigma, \delta, \rho)$, $t \in \langle \mathcal{A} \rangle$ and $\mathbf{x} \in \Sigma^*$ define $\mathrm{Orb}_t(\mathbf{x}) = \min_{\alpha > 0} \{\alpha \mid t^\alpha \cdot \mathbf{x} = \mathbf{x}\}$ the size of the orbit of $\mathbf{x}$ under the action of t. The *Orbit Signalizer* is the graph Γ_t whose vertices are the $t^{\mathrm{Orb}_t(\mathbf{x})}_{|\mathbf{x}}$, $\mathbf{x} \in \Sigma^*$ and edges from $t^{\mathrm{Orb}_t(\mathbf{x})}_{|\mathbf{x}}$ to $t^{\mathrm{Orb}_t(\mathbf{xy})}_{|\mathbf{xy}}$, $\mathbf{x} \in \Sigma^*, y \in \Sigma$ with label $\mathrm{Orb}_{t^{\mathrm{Orb}_t(\mathbf{x})}_{|\mathbf{x}}}(y)$.

The Orbit signalizer is used in [3,6] to solve the ORDER PROBLEM. Indeed the order of t is the lowest common multiple of all labels along paths starting from vertex t in Γ_t. In particular if the orbit signalizer is finite then the ORDER PROBLEM is decidable for t, as it reduces to checking if cycles have labels all 1.

3   Reversible Mealy Automata

We show that groups generated by invertible reversible Mealy automata have an element of infinite order with high probability. In fact we are going to prove a stronger result by showing that almost all invertible reversible automata are not bireversible, then using known results from [13], we obtain that the generated semigroups are almost surely torsion-free.

Since a Mealy automaton is completely defined by its transition and production functions, an invertible reversible Mealy automaton can be understood as $|Q|$ permutations in $S_{|\Sigma|}$ and $|\Sigma|$ permutations in $S_{|Q|}$, thus the uniform distribution on the set of invertible reversible Mealy automata with stateset Q and alphabet Σ is the uniform distribution on $S_{|Q|}^{|\Sigma|} \times S_{|\Sigma|}^{|Q|}$.

An invertible reversible automaton is bireversible if and only if each output letter induces a permutation of the stateset. In particular, for bireversible automata, we have that :

$$\forall (p, i), (q, j) \in Q \times \Sigma, (p, i) \neq (q, j), \delta_i(p) = \delta_j(q) \Rightarrow \rho_p(i) \neq \rho_q(j).$$

Define, for $r \in Q$, the set $\mathcal{O}_r = \left\{ j \in \Sigma, \exists (p, i) \in Q \times \Sigma, p \xrightarrow{i|j} r \in \mathcal{A} \right\}$ of output letters that lead to r. An invertible reversible automaton is bireversible if and only if, for all states r, the set $\mathcal{O}_r$ is the whole alphabet.

Example 3.1. Consider the automaton in Fig. 2. We have $\mathcal{O}_a = \{1, 2, 3\} = \mathcal{O}_c$ but $\mathcal{O}_b = \{1, 3\}$, hence the automaton is not bireversible.

Proposition 3.2. *The probability that a random invertible reversible automaton with k letters and n states is bireversible is less than*

$$\max\left\{\frac{1}{n^{k-1}} + \frac{1}{k}, \frac{1}{k^{n-1}} + \frac{1}{n}\right\}$$

Proof. Let $\mathcal{A} = (Q, \Sigma, \delta, \rho)$. For $r \in Q$, we denote $\mathrm{pred}_r = |\{p \in Q \mid \exists i \in \Sigma,\ \delta_i(p) = r\}|$ the size of the set of predecessor of r, and BIR the set of bireversible automata. Let us fix a state r. We have:

$$\mathrm{Pr}(\mathcal{A} \in \mathrm{BIR}) = \mathrm{Pr}(\forall q \in Q, \mathcal{O}_q = \Sigma)$$
$$\leq \mathrm{Pr}(\mathcal{O}_r = \Sigma).$$

From the law of total probability we get:

$$\mathrm{Pr}(\mathcal{A} \in \mathrm{BIR}) \leq \mathrm{Pr}(\mathcal{O}_r = \Sigma \mid \mathrm{pred}_r = 1)\,\mathrm{Pr}(\mathrm{pred}_r = 1)$$
$$+ \ \mathrm{Pr}(\mathcal{O}_r = \Sigma \mid \mathrm{pred}_r \geq 2)\,\mathrm{Pr}(\mathrm{pred}_r \geq 2)$$
$$\leq \mathrm{Pr}(\mathrm{pred}_r = 1) + \mathrm{Pr}(\mathcal{O}_r = \Sigma \mid \mathrm{pred}_r \geq 2)$$

The probability that r has exactly one predecessor can be seen as fixing $\delta_i^{-1}(r)$ for some reference letter $i \in \Sigma$ and requiring that the $k-1$ other $\delta_j^{-1}(r)$, $j \in \Sigma \setminus \{i\}$ are equal to $\delta_i^{-1}(r)$, hence:

$$\mathrm{Pr}(\mathcal{A} \in \mathrm{BIR}) \leq \frac{1}{n^{k-1}} + \mathrm{Pr}(\mathcal{O}_r = \Sigma \mid \mathrm{pred}_r \geq 2)$$

For the second term, let us consider a predecessor p of r and let λ be the number of input letters leading from p to r. We have $1 \leq \lambda \leq k-1$. To enforce bireversibility, we have to avoid that p outputs a letter that is already leading to r ($\rho_p(i) \neq \rho_q(j)$ for $\delta_p(i) = \delta_q(j) = r$). Assume that the set $\mathcal{O}_r^{(p)} = \left\{ j \in \Sigma \mid \exists q \neq p \in Q, q \xrightarrow{i|j} r \in \mathcal{A} \right\}$ of output letters leading to r from a state q different from p is of maximal size $k - \lambda$. Since ρ_p is random and independent from the others ρ_q we can bound this probability from above: having the letters leading from p to r produce the λ out of k required letters is $\binom{k}{\lambda}^{-1}$. Hence:

$$\mathrm{Pr}(\mathcal{A} \in \mathrm{BIR}) \leq \frac{1}{n^{k-1}} + \binom{k}{\lambda}^{-1}$$
$$\leq \frac{1}{n^{k-1}} + \frac{1}{k}.$$

Now, by applying the same reasoning to the dual automaton, which is invertible and reversible on k states and n letters, we get the symmetric upper bound

$$\mathrm{Pr}(\mathcal{A} \in \mathrm{BIR}) \leq \frac{1}{k^{n-1}} + \frac{1}{n}.$$

$\square$

It is proven in [13] that an invertible reversible Mealy automaton without bireversible component generates a torsion-free semigroup[1]. Whence our theorem:

Theorem 3.3. *The probability that an invertible reversible Mealy automaton taken uniformly at random generates a torsion-free semigroup goes to 1 as the size of the alphabet grows. Moreover, the probability for the group to have an element of infinite order also goes to 1 as the stateset or the alphabet grows.*

Proof. It is known that, with great probability, two random permutations on a large set generate a transitive group [8]: more precisely, he proved that two random permutations of S_k generate a transitive group with probability $1 - 1/k + O(1/k^2)$. In terms of a graph, it means that a typical reversible Mealy automaton on a large alphabet is (strongly) connected, and is not bireversible by Proposition 3.2 whence the first part of the result. The second part is from Proposition 3.2 and [13]. □

From [9], where it is shown that having an element of infinite order implies exponential growth among groups generated by invertible reversible automata, we obtain:

Theorem 3.4. *The probability that an invertible reversible Mealy automaton taken uniformly at random generates a group with exponential growth goes to 1 as the size of the stateset or of the alphabet grows.*

Remark 3.5. Notice that Theorem 3.3 is not *a priori* a consequence of Theorem 3.4: there exists infinite Burnside group with exponential growth ([1]). However, no example of such a group is known within the class of automaton groups.

It is worthwhile noting that it is unknown whether the ORDER PROBLEM is decidable within the class of (semi)groups generated by reversible Mealy automata.

4   Reset Mealy Automata

The class of reset automata is of particular interest since it is linked to *one-way cellular automata*, and was used by Gillibert to prove the undecidability of the ORDER PROBLEM for automaton semigroups [10]. For groups generated by (invertible) reset Mealy automata the ORDER PROBLEM remains open [7].

As the transition function is trivial in a (unfolded) reset Mealy automaton, the uniform distribution on the set of unfolded invertible reset Mealy automata with stateset Q and alphabet Σ is the uniform distribution on $S_{|\Sigma|}^{|Q|}$.

We are going to use a result from [20]:

---

[1] Notice that an invertible Mealy automaton might generates a torsion-free semigroups but a group which is not torsion free. For instance the classical lamplighter group generated by a bireversible automaton which generates a torsion-free semigroup [13,16].

Theorem 4.1 ([20, Theorem 1.20]). *Let $\mathcal{A}$ be a reset automaton and $\pi_{\mathcal{A}}$ be the transformation defined by $\pi_{\mathcal{A}} : q \mapsto \rho_q^{-1}(q)$ for all $q \in Q$.. If $\pi_{\mathcal{A}}$ is not a permutation then the group generated by $\mathcal{A}$ has an element of infinite order.*

We give the proof for the sake of completeness.

Proof. If $\pi_{\mathcal{A}}$ is not a permutation, then there exists x_0 which does not belong to any cycle of $\pi_{\mathcal{A}}$ and such that $\pi_{\mathcal{A}}(x_0) = x_1$ belongs to a cycle $x_1 \to \cdots \to x_\ell \to x_1$ of $\pi_{\mathcal{A}}$. Computing the orbit of $x_0(x_1 \cdots x_\ell)^{\alpha}$ under the action of any given state $q \in Q$ gives:

$$
\begin{array}{ccccccccccc}
& x_0 & & x_1 & & x_2 & \cdots & & x_\ell & & x_1 \cdots \\
q & \xrightarrow{} & x_0 & \xrightarrow{} & x_1 & \xrightarrow{} & x_2 \cdots x_{\ell-1} & \xrightarrow{} & x_\ell & \xrightarrow{} & x_1 \\
& y_1 & & x_0 & & x_1 & \cdots & & x_{\ell-1} & & x_\ell \\
q & \xrightarrow{} & y_1 & \xrightarrow{} & x_0 & \xrightarrow{} & \cdots x_{\ell-2} & \xrightarrow{} & x_{\ell-1} & \xrightarrow{} & x_\ell \\
& y_1' & & y_2' & & x_0 & \cdots & & x_{\ell-2} & & x_{\ell-1} \\
& \vdots & & \vdots & & \vdots & & & \vdots & & \vdots
\end{array}
$$

Since $x_1 = \rho_{x_0}^{-1}(x_0) = \rho_{x_\ell}^{-1}(x_\ell)$. So $q^{\alpha i} \cdot x_0 (x_1 \cdots x_\ell)^{\alpha} = \mathbf{u} x_0 (x_1 \cdots x_\ell)^{(\alpha-i)}$, for some $\mathbf{u} \in Q^{i\alpha}$, hence q has infinite order. $\qquad\square$

Theorem 4.2. *The probability that a random (unfolded) reset automaton on k letters has an element of infinite order is at least $1 - e\sqrt{k}e^{-k}$.*

Proof. Since the ρ_q, $q \in \Sigma$ are random permutations, the function $q \mapsto \rho_q^{-1}(q)$ can be considered as a random mapping from Σ to Σ, and the number of permutations among mappings is $\frac{k!}{k^k}$. We conclude using Stirling's approximation and more precisely [21]: $k! < \sqrt{2\pi k}(\frac{k}{e})^k e^{\frac{1}{12k}} < e\sqrt{k}k^k e^{-k}$ as soon as $k > 1$. $\qquad\square$

Using [19], where Olukoya proves that groups generated by reset automata are either finite or have exponential growth, we get (see also Remark 3.5):

Theorem 4.3. *The probability that a random (unfolded) reset automaton on k letters has exponential growth is at least $1 - e\sqrt{k}e^{-k}$.*

From Delacourt and Ollinger [7, Proposition 1], our result also means that permutive one-way cellular automata are generically aperiodic.

Remark 4.4. An unfolded reset automaton is minimal ([2]) if and only if each state induces a different permutation on letters. By the birthday problem, we can extend our result a bit: a random minimal unfolded reset automaton generically generates a group with exponential growth and elements of infinite order.

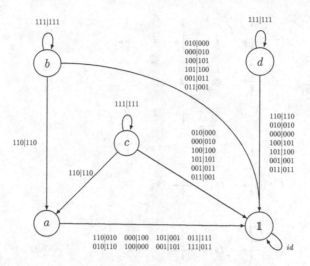

Fig. 3. The normal form $^3\mathcal{G}$ of the (invertible) automaton $\mathcal{G}$ of bounded activity generating the Grigorchuk group. A simple path contains at most one nontrivial loop.

5   Mealy Automata with Polynomial Activity

The class of Mealy automata with polynomial activity is interesting as the ORDER PROBLEM is decidable for (semi)groups generated by automata with bounded activity but remains open for the higher levels of the hierarchy [6].

Recall that $\mathbb{1}$ denotes the identity state in the automaton, which is supposed to be unique. We are going to define a *normal form*: let $\mathcal{A} = (Q, \Sigma, \delta, \rho)$ be an automaton with polynomial activity and let ℓ be the lowest common multiple of the sizes of the (simple) cycles. Since an automaton with polynomial growth has no entangled cycles, we have that $^\ell\mathcal{A}$ has all cycles of length one. Now put d the maximal length of an (oriented) path between a state and a self-loop in $^\ell\mathcal{A}$. Then the normal form of the automaton $\mathcal{A}$ is the automaton $^{d\ell}\mathcal{A}$ and it looks as follows: it is a directed acyclic graph whose leaf induces the identity $\mathbb{1}$, and where each state either has a self-loop or leads to a state with a self-loop. For instance the normal form $^3\mathcal{G}$ of the (invertible) Grigorchuk automaton $\mathcal{G}$ (which has bounded activity) Fig. 1 (left) is depicted Fig. 3.

To the best of our knowledge, and even among automata under normal form, there is no easy description of the uniform distribution on the set of (invertible) Mealy automata with polynomial activity, even if one fixes the degree of the activity. To bypass this difficulty, we show that automata with finitary activity are rare even among automata with bounded activity, and use the fact that once the transition functions are fixed, the choice of production functions does not change the activity.

The next proposition is a simple yet useful observation:

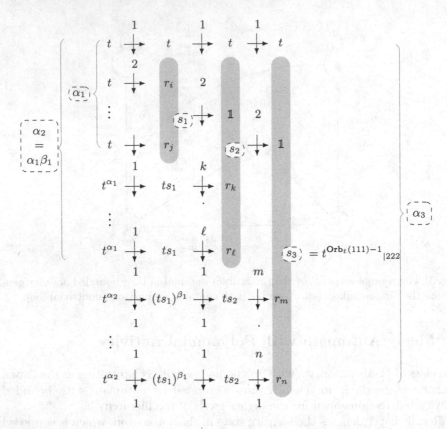

Fig. 4. Computation of the orbit of 1^j under the action of t in a bounded automaton.

Proposition 5.1. *Let $\mathcal{A} = (Q, \Sigma, \delta, \rho)$ be a Mealy automaton with bounded activity under normal form. If there is some $t \in Q$ with $\delta_i(t) = t$ and $\rho_t(i) \neq i$, then ρ_t has infinite order.*

Proof. Up to renaming, we can assume that $i = 1$ and $\rho_t(1) = 2$. We use the orbit signalizer of t to prove that ρ_t has infinite order (see Fig. 4): put $\alpha_j = |\mathrm{Orb}_t(1^j)|$. Since the activity is bounded, the set $\left\{ t^{|\mathrm{Orb}_t(1^j)|}_{|1^j} \right\}_j$ is finite, so there is a self-loop $t^{\alpha_i}_{|1^i} = t^{\alpha_{i+j}}_{|1^{i+j}}$. By putting $t^{\alpha_i}_{|1^i} = ts_\alpha$ we get that $(ts_i)^\beta_{|1^j} = ts_i$ for some integer $\beta = |\mathrm{Orb}_{ts_i}(1^j)|$. Suppose that the size of the orbit is 1. We obtain:

$$
\begin{array}{ccc}
& 1^j & \\
t \xrightarrow{} t & \quad & \\
\downarrow 2^j & \text{but} & \\
s_i \xrightarrow{} s_i & & \\
\downarrow 1^j & &
\end{array}
\qquad
\begin{array}{c}
1 \\
t \xrightarrow{} t \\
\downarrow 2 \\
s_i \xrightarrow{} \mathbb{1} \\
\downarrow 1
\end{array}
\qquad \text{so } s_i = \mathbb{1} \text{ and}
\qquad
\begin{array}{c}
1 \\
t \xrightarrow{} t \\
\downarrow 2 \\
\mathbb{1} \xrightarrow{} \mathbb{1} \\
\downarrow 1
\end{array}
$$

Hence $2 = 1$, contradiction. The size of the orbit of ρ_t increases strictly through the cycle 1^β, whence the order of ρ_t–the lowest common multiple of the paths in the orbit signalizer of t ([6])–is infinite.   □

From this result, we get:

Theorem 5.2. *The probability that the group generated by an automaton with polynomial activity has an element of infinite order goes to 1 as the size of the alphabet goes to infinity.*

Proof. We first prove that groups generated by automata with polynomial non finitary activity generically have an element of infinite order: let $\mathcal{A} = (Q, \Sigma, \delta, \rho)$ be an automaton in $\mathrm{Pol}(d) \setminus \mathrm{Pol}(-1)$ and let t be a state with bounded activity on a nontrivial cycle. Since the activity does not depend on the choice of the production functions (except for the trivial state), we can consider the set $\mathcal{C}_\mathcal{A}$ of automata in $\mathrm{Pol}(d) \setminus \mathrm{Pol}(-1)$ with same transition functions and trivial state. Among $\mathcal{C}_\mathcal{A}$, we have $\rho_t(i) \neq i$ with probability $1 - 1/k$, so, in the normal form, t is on a cycle labelled by $i\mathbf{x} \in \Sigma^\ell$ with $\rho_t(i\mathbf{x}) \neq i\mathbf{x}$. We can apply Proposition 5.1.

Now we show that the set $\mathrm{Pol}(-1)$ has measure 0 in the set $\mathrm{Pol}(d)$, $d \geq 0$. If an automaton $\mathcal{A}$ has polynomial activity, then there is at least one state t satisfying $\delta_i(t) = \mathbb{1}$ for all i. Given $\mathcal{A} \in \mathrm{Pol}(-1)$, we can build k automata $\mathcal{A}_i$ with bounded but not finitary activity by changing for exactly one letter $\delta_i(t) = \mathbb{1}$ to $\delta_i(t) = t$. If we consistently chose t to be, *e.g.* , the minimal among acceptable states, we can uniquely reconstruct $\mathcal{A}$ from these $\mathcal{A}_i$, whence the result.
We conclude using the law of total probability: the probability that an automaton in $\mathrm{Pol}(d)$ has an element of infinite order is equal to the probability that it has an element of infinite order given it belongs to $\mathrm{Pol}(d) \setminus \mathrm{Pol}(-1)$ times the probability of the later; we showed that both go to one, the result follows.   □

From the proof we extract the following:

Proposition 5.3. *The probability that the group generated by an automaton in $\mathrm{Pol}(0)$ on an alphabet of size k has an element of infinite order is at least $\frac{k-1}{k+1}$.*

6   Conclusion and Future Work

In this work, we proved, for various important classes of Mealy automata, that the generated groups have generically an element of infinite order, thus are infinite. It is natural to wonder whether other properties, such as non-amenability, are generic and to extend these results to the full class of automaton groups.

One interesting direction is to determine if generating a free or an infinitely presented group is generic in a class. These properties are mutually exclusive. Automata with polynomial activity cannot generate free groups [18], while reversible ones can [25]; infinitely presented groups can be found in both classes. It would be striking to find two classes and a group property which is nontrivial in both classes yet generically true in one and generically false in the other.

Acknowledgements. The author thanks Ville Salo who asked the question that initiated this work and for interesting discussions, and Matthieu Picantin and Jérémie Brieussel for their comments during the redaction of this paper. He also thanks the anonymous reviewers for their remarks which helped to improve the legibility of the paper. He was supported by Academy of Finland grant 296018 and by the French *Agence Nationale de la Recherche* through the project AGIRA.

References

1. Adian, S.I.: The Burnside problem and identities in groups, Ergebnisse der Mathematik und ihrer Grenzgebiete **95** (1979)
2. Akhavi, A., Klimann, I., Lombardy, S., Mairesse, J., Picantin, M.: On the finiteness problem for automaton (semi)groups. Int. J. Algebr. Comput. **22**(6), 1–26 (2012)
3. Bartholdi, L., Godin, T., Klimann, I., Picantin, M.: A new hierarchy for automaton semigroups. In: Câmpeanu, C. (ed.) CIAA 2018. LNCS, vol. 10977, pp. 71–83. Springer, Cham (2018). https://doi.org/10.1007/978-3-319-94812-6_7
4. Bartholdi, L., Mitrofanov, I.: The word and order problems for self-similar and automata groups (2017). arXiv:1710.10109
5. Bartholdi, L., Virág, B.: Amenability via random walks. Duke Math. J. **130**(1), 39–56 (2005)
6. Bondarenko, I.V., Bondarenko, N.V., Sidki, S.N., Zapata, F.R.: On the conjugacy problem for finite-state automorphisms of regular rooted trees. Groups Geom. Dyn. **7**(2), 323–355 (2013). With an appendix by R. M. Jungers. MR 3054572
7. Delacourt, M., Ollinger, N.: Permutive one-way cellular automata and the finiteness problem for automaton groups. In: Kari, J., Manea, F., Petre, I. (eds.) CiE 2017. LNCS, vol. 10307, pp. 234–245. Springer, Cham (2017). https://doi.org/10.1007/978-3-319-58741-7_23
8. Dixon, J.D.: The probability of generating the symmetric group. Mathematische Zeitschrift **110**(3), 199–205 (1969)
9. Francoeur, D., Mitrofanov, I.: On the existence of free subsemigroups in reversible automata semigroups (2018). arXiv:1811.04679
10. Gillibert, P.: The finiteness problem for automaton semigroups is undecidable. Int. J. Algebr. Comput. **24**–1(1), 1–9 (2014)
11. Gillibert, P.: An automaton group with undecidable order and Engel problems. J. Algebr. **497**, 363–392 (2018)
12. Godin, Th.: An analogue to Dixon's theorem for automaton groups. In: Proceedings of the Fourteenth Workshop on Analytic Algorithmics and Combinatorics (ANALCO), pp. 164–173 (2017)
13. Godin, T., Klimann, I., Picantin, M.: On torsion-free semigroups generated by invertible reversible mealy automata. In: Dediu, A.-H., Formenti, E., Martín-Vide, C., Truthe, B. (eds.) LATA 2015. LNCS, vol. 8977, pp. 328–339. Springer, Cham (2015). https://doi.org/10.1007/978-3-319-15579-1_25
14. Grigorchuk, R.I.: On Burnside's problem on periodic groups. Akademiya Nauk SSSR. Funktsional′nyĭ Analiz i ego Prilozheniya **14**–1, 53–54 (1980)
15. Grigorchuk, R.I.: Degrees of growth of finitely generated groups and the theory of invariant means. Izvestiya Akademii Nauk SSSR. Seriya Matematicheskaya **48**–5(5), 939–985 (1984)
16. Grigorchuk, R.I., Nekrashevich, V.V., Sushchanskiĭ, V.I.: Automata, dynamical systems, and groups. Trudy Matematicheskogo Instituta Imeni V. A. Steklova. Rossiĭskaya Akademiya Nauk **231**, 134–214 (2000)

17. Nekrashevych, V.: Self-Similar Groups. Mathematical Surveys and Monographs, vol. 117. American Mathematical Society, Providence (2005)
18. Nekrashevych, V.: Free subgroups in groups acting on rooted trees. Groups Geom. Dyn. 4(4), 847–862 (2010)
19. Olukoya, F.: The growth rates of automaton groups generated by reset automata (2017). arXiv:1708.07209
20. Picantin, M.: Automates, (semi)groupes, dualités, Habilitation á diriger des recherches, Université Paris Diderot (2017)
21. Robbins, H.: A remark on Stirling's formula. Am. Math. Mon. 62(1), 26–29 (1955)
22. Russyev, A.: Finite groups as groups of automata with no cycles with exit. Algebr. Discret. Math. 9(1), 86–102 (2010)
23. Sidki, S.: Automorphisms of one-rooted trees: growth, circuit structure, and acyclicity. J. Math. Sci. 100(1), 1925–1943 (2000). Algebra, 12
24. Silva, P., Steinberg, B.: On a class of automata groups generalizing lamplighter groups. 15, 1213–1234 (2005)
25. Vorobets, M., Vorobets, Y.: On a free group of transformations defined by an automaton. Geometriae Dedicata 124, 237–249 (2007)
26. Wilson, J.S.: On exponential growth and uniformly exponential growth for groups. Invent. Math. 155(2), 287–303 (2004). MR 2031429

Author Index

Printed in the United States
By Bookmasters